Islam Obscured

The Rhetoric of Anthropological Representation

Daniel Martin Varisco

First published in 2005 by
PALGRAVE MACMILLAN™
175 Fifth Avenue, New York, N.Y. 10010 and
Houndmills, Basingstoke, Hampshire, England RG21 6XS
Companies and representatives throughout the world.

PALGRAVE MACMILLAN is the global academic imprint of the Palgrave Macmillan division of St. Martin's Press, LLC and of Palgrave Macmillan Ltd. Macmillan® is a registered trademark in the United States, United Kingdom and other countries. Palgrave is a registered trademark in the European Union and other countries.

ISBN 1–4039–6772–5
ISBN 1–4039–6773–3

Library of Congress Cataloging-in-Publication Data

Varisco, Daniel Martin.
    Islam obscured : the rhetoric of anthropological representation / Daniel Martin Varisco.
        p. cm.—(Contemporary anthropology of religion)
    Includes bibliographical references (p.) and index.
    ISBN 1–4039–6772–5 (hc : alk. paper)
    ISBN 1–4039–6773–3 (pbk.: alk. paper)
        1. Islam. 2. Ethnology—Islamic countries. 3. Islam and culture—Islamic countries. I. Title. II. Series.

BP161.3. V37 2005
306.6'97—dc22   2004050514

A catalogue record for this book is available from the British Library.

Design by Newgen Imaging Systems (P) Ltd., Chennai, India.

First edition: January 2005

10 9 8 7 6 5 4 3 2 1

Printed in the United States of America.

ISLAM OBSCURED

CONTEMPORARY ANTHROPOLOGY OF RELIGION

*A series published with the Society for the
Anthropology of Religion*

Robert Hefner, Series Editor
Boston University
Published by Palgrave Macmillan

Body / Meaning / Healing
By Thomas J. Csordas

*The Weight of the Past: Living with History in Mahajanga,
Madagascar*
By Michael Lambek

*After the Rescue: Jewish Identity and Community in Contemporary
Denmark*
By Andrew Buckser

Empowering the Past, Confronting the Future
By Andrew Strathern and Pamela J. Stewart

The Muslim should learn to look more objectively at his religious history, particularly at how Islam has fared at his hands, and the non-Muslim should learn to know something of what Islam does to a Muslim from the inside.

Fazlur Rahman

The anthropologist taking a phenomenological approach focuses on the daily lived experience of the local islams and leaves the study of theological interpretation to the Islamists.

Abdul Hamid el-Zein

Contents

Acknowledgments

This book has evolved over several years of deciding what it should be. Along the way a number of colleagues have patiently responded to my incessant emails and, at times, graciously given of their time to read drafts of chapters in the book. Among these are Najwa Adra, Jon Anderson, Dale Eickelman, Steve Caton, Andre Gingrich, Terry Godlove, Bob Hefner, David Hicks, Sharryn Kasmir, Kathy Kueny, Ron Lukens-Bull, Rich Martin, Laury Silvers, Ira Singer, Richard Tapper, Shelagh Weir, and Bill Young. Many Muslim friends and students have helped shape my thinking for this book in ways they would probably not recognize.

After finishing the draft of this book, I received news of the passing of George Makdisi, a distinguished historian who had served on my doctoral committee at the University of Pennsylvania. Over the years he was a mentor to me on matters philological and pragmatic. Although this book does not touch on the subjects of Islamic law and education to which he made notable contributions, I do owe part of my abiding penchant for critical engagement to the demanding standards Makdisi set for himself as well as for his students. Mindful of his influence, which never allows me to be comfortable with established thinking, I dedicate this book to George Makdisi, an individual who, in my mind, defines what it means to be a scholar and a gentleman.

The chapter on Geertz was given in roughly its present form at the annual meeting of the Society for the Anthropology of Religion in Cleveland, Ohio, in April, 2002. The chapter on Mernissi stems from a paper delivered at the American Anthropological Association in Washington, D.C., 1995. I delivered the chapter on Ahmed at the annual meeting of the Society for the Anthropology of Religion in Providence, Rhode Island, in April, 2003. I wish to thank Yale University Press for permission to quote from Geertz's *Islam Obscured*.

Introduction

Anthropology and Islam

But to conceptualize Islam as an object of an anthropological *study is not as simple as some writers would have one to suppose.*[1]

Talal Asad

What the world does not need is yet another book that assumes Islam can be abstracted out of evolving cultural contexts and neatly essentialized into print without repeating the obvious or glossing over the obtuse. This is—I believe and I hope—not such a book. I have no interest in telling you what Islam *is*, what it really *must be*, or even what it *should be*. In what follows I am more attuned to what Islam hopefully is not, at least not for someone who approaches it seriously as an anthropologist and historian. I bear no obvious axe to grind as either a determined detractor against the religion or an overanxious advocate for it. Personally, as well as academically, I consider Islam a fascinatingly diverse faith, a force in history that must be reckoned with in the present. The offensive tool I do choose to wield, if my figurative pen can stand a militant symbol, is that of a critical hammer, an iconoclastic smashing of the rhetoric that represents, overrepresents, and misrepresents Islam from all sides. By avoiding judgment on the sacred truth of this vibrant faith, I shift intention toward an I-view that takes no summary representation *of* Islam as sacred.

Like any revelation that expects to be taken seriously, Islam is about truth in all its various forms. It has become fashionable in the postexistential, poststructural, post-colonial, and temporarily postmodern climate of much intellectual criticism to ignore truth claims, reducing them to mere representation or simply by sinking into the quagmired once-metaphysical debate over what truth could possibly mean. Nietzsche is not my theoretical niche; nor do I wish to follow Foucault into self-contained deciphering of discourse or Derrida down the deconstructive path of linguistic relativism. Although I have no meta-truth to reveal, neither do I smugly assume that Islam is not or could not be true in the experiential sense knowable only to a believer. For Muslims the truth is best seen, as Fazlur Rahman wisely suggests, from the inside.[2] As an anthropologist, I am prepared to follow Abdul Hamid el-Zein and leave such verification of truth to the theologians.[3] My concern at the offset is with the outside, the rhetoric of representing Islam as a religion through the lens of anthropological or sociological narratives. Much of what has been written and is still sadly said, with academic air as well as media flair, is so overflowing with half truths and untruths wrapped around grains of truths that the dynamics of one of the world's largest and fastest growing religions are obscured.

There are two major reasons for writing this book. The first is academic, and hopefully more than academic: there is no up-to-date, critical assessment of how anthropologists have represented the religion of Islam. A century after the demise of mainstream Christian missionary apologetics against its Oriental rival and just a few decades into an academic evolution beyond the kind of old-style "Orientalism" savaged by Edward Said, the study of Islam in the widest sense stretches across various disciplines and post-ed counters. For somewhat less than a century ethnographers have observed "Islam" where it is practiced; there is now a wide and relevant corpus of ethnographic data and analysis available. Although certainly not the dominant voices representing Islam, anthropologists and sociologists today figure in the process because of what they are able to learn by observing the behavior and rhetoric of Muslims in social contexts, usually in non-Western societies. Yet few scholars outside anthropology, as well as many within the general ranks, are aware of the ways in which the rhetoric in this corpus has changed against the backdrop of postmodern critique of ethnography as a genre and the shifting paradigms within the field. Contemporary anthropology is not the exotica and erotica trope that so many people assume it always used to be.

So why not write an intellectual history of the anthropological study of Islam in order to say who did what, when, where, and how (I prefer to leave the why to the psychoanalysts)? This was my initial impulse: compile a comprehensive bibliography, plot the trajectory through specific ethnographic texts and theoretical discussions, end up with a 700-page tome that only a few well-endowed libraries would buy. Such a project, were I ever to return to it, could easily become one of those always-looking-for-something-I-missed stories that never ends. Instead I decided to return to a few overarching anthropological treatments of Islam. This led me to choose these four authors and their seminal texts. Rather than survey what anthropologists of varying persuasions have said about Islam or summarize the gists for student consumption, I decided to contextualize these texts based not only on reading of relevant ethnographies but also from my own ethnographic experience and historical research on various aspects of Islamic traditions in the Middle East.

Several post-participation observers have returned from the field to offer suggestions on how anthropology could or should treat Islam, but these must be fleshed out of ethnographic monographs, journal articles, and extended book reviews. Surprisingly, only a few anthropologists have been tempted to propose a way of looking at Islam in the cultural aggregate. The seminal text—widely recognized both within and without the discipline of American anthropology—that stands for an "anthropology" of Islam remains Clifford Geertz's far-too-well-traveled *Islam Observed* (1968). Across the Atlantic the main anthropological/sociological theorist of Islam has been Ernest Gellner, who in *Muslim Society* (1981) weaves his model of representation out of a many-colored philosophical cloak with strands from David Hume, Max Weber, Ibn Khaldun, and a host of Englightenment mentors. The perspective of a Muslim feminist was provided to Western readers by Moroccan sociologist Fatima Mernissi, whose *Beyond the Veil* (1975) is one of the first "sociological" analyses of Muslim gender roles. Another Muslim scholar, the British/Pakistani anthropologist Akbar Ahmed, looks at his own religion inside out in *Discovering Islam* (1988), building in principle on an avowedly "Islamic" mode of anthropology. These are not the only English-language studies presentable under the rubric of an "anthropology of Islam," but I believe that they have been the most widely read and consulted in the latter part of the twentieth century.[4] Regardless of their age, all remain in active print and prominent library use in the early measure of the twenty-first century. As an anthropologist reading and rereading other anthropologists, I offer here a critique of the rhetoric of representing Islam in the texts of Geertz, Gellner, Mernissi, and Ahmed.

The reason to engage in such a critique is, as hinted above, more than academic. Textual truths engendered and far too often engineered in representing Islam find their way unscrutinized and insufficiently digested into an endless stream of introductory and general texts, even solidly scholarly works. Seminal texts, once canonized as theoretically innovative or simply authoritative by default, have a library shelf life far beyond their usefulness and freshness in the disciplines that generate them. This is by no means a dilemma specific to anthropology, but it is a problem that needs to be addressed and mitigated, even if only one discipline at a time. Contemporary scholars not schooled in anthropological theory at the graduate level should no longer be excused for thinking that anthropology's "classics" written

by Ruth Benedict, Margaret Mead, Robert Redfield, and—yes—
Clifford Geertz still resonate as distinctly in the critical advances of
recent anthropological thinking. It is not that older texts get proven
wrong so much as that they simply run out of inspirational steam for
the stream of new students and the stamina of those in the old guard
willing to think beyond tenure. Seldom does a new book come along
and get filed next to the old stalwarts; publishers routinely reprint
classics with no-fault insurance against critical journal reviews. The
very nature of any scholar traveling, trickle that this probably is, to a
discipline like anthropology for theoretical insights is thus fraught
with pitfalls when the analysis available is increasingly measured
backward by the decade.

Scholars who study religion often refer to the great tradition/little
tradition trope of vintage 1950s Robert Redfield, a prominent
anthropologist at the University of Chicago. I suspect that few who
apply Redfield's seemingly unforgetable notion of great and little
traditions are aware of how it came about. He recognized, quite
rightly and timely for the mid-twentieth century, that the prevailing
anthropological concept of distinct cultures owed much to the
emphasis on isolated tribes and peoples that could conveniently
be buried in cultural wholes. Unlike a number of well-known ethnographers who went off deep into the thickly descriptive jungles or
academically marooned themselves on exotic isles, Redfield focused
his attention on Mexican peasants. "The culture of a peasant community, on the other hand," reasoned Redfield from experience, "is not
autonomous."[5] Life in the village he lived in was not self-contained,
even in principle. It was not enough to just study behavior in the
village because of the numerous intellectual, political, and economic
links to the wider civilization. Redfield proposed a way "to take
mental hold of this compound culture," specifically the difference
between a great tradition outside the village and a little tradition
observable at the local level. "In a civilization there is a great tradition of
the reflective few," he writes famously, "and there is a little tradition of
the largely unreflective many."[6] This was hardly a novel concept, as
Redfield himself notes, but simply a recognition that ethnography in
literate, especially Western, contexts needed to be more than "being
there" among not-as-exotic others.

After Redfield introduced the idea, the historian of Islam, Gustave
von Grunebaum, a Chicago colleague of Redfield, latched onto this
simple bifurcation, perhaps under the self-fulfilling impression that
the "great" stuff was reserved for trained Arabists and historians while
the "little" bits were suitable leftovers for folklorists and anthropologists.[7] Redfield was pleased that the "Islamist comes to meet the

anthropologist," a melding of the view from the top with the view from the trenches.[8] But, lacking ethnographic experience in Islamic contexts, he was unable to check the prevailing academic *faux pas* of an unchanging Orient as an overlay of historical real time. "From what I read," observes Redfield naively, "the Magreb of Morocco even today provide [*sic*] an instance of an ancient and little changed structure of Islamic sacred tradition."[9] Thus, he continues, "there is practically no difference there between a work written in the sixteenth century and one written in the twentieth, newspapers are unknown, and the intellectual life is confined to a small elite who are concerned ever with the same problems of interpreting Muslim orthodoxy." Such is the lamentable result of a pre-Geertzian Islam unobserved.

Redfield's reflections have a metaphorical elegance, and certainly the intention was right in stressing the need to integrate the local with ever-expanding worlds of meaning. "Great and little traditions can be thought of as two currents of thought and action, distinguishable, yet ever flowing into and out of each other," he suggests. As an example, he notes that the Quran "has the content it has because it arose among Arab not Chinese peoples," and the teachings established as a "great" tradition are not necessarily those held by generations of peasants. These are self-evident points that virtually any scholar today would accept. Yet, one problem with Redfield's elaboration is that it reifies an artificial distinction between civilized and primitive societies. Some societies, but not the ones he studied, are said to have no "great tradition." By default, any given people will weigh in on the scale of civilization according to how literate and reflective they are. Thus, certain practices by Muslims are denigrated as "superstition" because they do not conform with an alleged "great" ideal.[10] Further problems arise in the distinguishing: who is to say which is greater, how great does a tradition have to be in order not to be little anymore? A great deal of new thinking has come along in the intervening years since this model was first proposed and then canonized. Unfortunately the dichotomy as such has perpetuated a profound Western ethnocentrism.[11]

By the mid-1970s, the Moroccan historian Abdallah Laroui fixed on historian von Grunebaum, the "Islamist," as paradigmatic for how "cultural anthropology" of Arabs has been done.[12] Although von Grunebaum had no training in the discipline and no experience in the field, the cultural anthropology attributed to him was summarily dismissed strawman style. Neither von Grunebaum nor anthropologist Redfield should serve as relevant examples for the ethnographic study of Muslims, a study barely on the map at the time they wrote. Laroui can be forgiven for not being aware of the contemporaneous rejection of this great/little binary by anthropologists studying Islam in the field.[13] Abdel Hamid el-Zein, an ethnographer who had conducted research in a community on the Kenyan island of Lamu, argued about the same time as Laroui that dividing Islam into great/little or formal/folk was an illusion. In any cultural context, argued el-Zein, "a folk theology may be found which rivals formal theology in its degree of abstraction, systematization, and cosmological implication."[14] But the momentum of assuming that anthropology can be summed up by one notion or one individual is not easily halted; even a prominent historian of religion assumes the working model of great/little Islam, in all innocence, two decades later.[15]

Another compelling motivation for a book in the format of textual critique is the need to speak out to colleagues and the general public about the continuing reprehensible representations of Islam and Middle Eastern people in Western society at large and in the news media. Deeply rooted ethnocentric prejudice and an unwillingness to see beyond political expedience have contributed to a demonization of Islam as a religion of violent terror alongside the older Judaeo-Christian charge of heretical error. Several prominent media icons of the Christian right have gone so far as to label the prophet Muhammad a "terrorist" and the Quran as the "enemy's book."[16] Recent collective cultural memories, whether premeditated or self-mediated, comprise an inescapably politicized litany: oil embargo, hostage crisis, mad mullahs, shoe-string budgeted airplane hijackings, skyscraper terrorism, Hamas suicide bombers, and the uncivilized clash with a post-red, green menace of fundamentalist militants. The cycle of blaming victims and victimizers, from CNN crossfiring to talk-radio jockeying and Internet chat rooms, ensures that "Islam" will be viewed suspiciously as a "problem" by Americans and Europeans for the foreseeably intolerant future.

What went wrong? How did the ideologically driven politics of nationalism and neocolonial birth pangs lose out in causal terms to the rantings of religious extremists and the martyrdom of children? The pundits have mostly played a blaming game. Echoing the patriotic rhetoric of President George W. Bush after 9/11, historian Bernard Lewis traces the troubles of the Islamic world to a "lack of freedom," seemingly the failure of predominantly Muslim countries to have the same governmental ancestry as France or the United States.[17] Were Muslims secular in a Western mode, the argument implies, they too could become enlightened enough to reform their religion into irrelevance. For a political scientist like Martin Kramer, the sandtrap lies with the entire Middle East Studies establishment reinvented in the wake of Edward Said's critique of media-friendly establishment scholars like Bernard Lewis. Middle East specialists who recognized and lamented the ethnocentrism and racism of an imperially aligned "Orientalism" are faulted for a "failure to anticipate

Islamism."[18] Being a blindsided expert on Islam in this scenario becomes tantamount to being a geologist who fails to predict the timing of an earthquake or a stockbroker who does not foresee a recession. Op-ed speak aside, there was hardly a need for academic scholars studying Islam—certainly not those who have no expertise as "political" scientists—to predict the swell of political unrest couched in religious rhetoric throughout the Muslim world. No one, not even the most sophisticated intelligence operatives in the world, was able to predict the attack on the Twin Towers. Were American security advisors waiting for an Ivory Tower directive to tell them the obvious: American policy toward the Middle East has continually generated violent reactions? Asking what went wrong in order to trounce one's opponents is disingenuous; the real question should be why ongoing global power plays resulting in political instability, economic disparity, cultural defamation, and misplaced self-interest should be labeled failures of religion. What went wrong is what usually goes wrong: someone else gets the blame for not being on the right side of God.

Frustration over political and economic events has mired much representation of Islam, the religion, into a referendum on cultural difference. Uncovering the inescapable truths about what—as Edward Said some time ago most forcefully brought to the public's attention—happens in "covering" Islam is not yet a done deed. There is, fortunately, fair and objective commentary on various aspects of Islam, even if one must first sort through the blatantly biased accounts and recycled rubbish of commercially littered books on "Islam" and "Arab" in the post-9/11 publishing world. Most of this coverage, especially the newstand variety, has been little influenced by what anthropologists have observed in the behavior and speech of ordinary Muslims. Nor has the rhetoric of readily available anthropological texts, when consulted, been subjected to sustained critique from within the ranks of anthropologists with field experience. This makes it all the more important to know both how anthropologists have represented Islam and what more could be said based on the potential of ethnographic research and comparative cultural analysis. I acknowledge a major failure of anthropological writers, apart from those who texts I examine, to reach a broad audience. *Inshallah*, this book is a step in the right direction.

Where Islam Comes From

It is not to distress you that We revealed the Koran, but to admonish the God-fearing. It is a revelation from Him who has created the earth and the lofty heavens, the Merciful who sits enthroned on high.[19]

Surah *Ta Ha*, Quran

For time-conscious historians, the Islamic revelation entered world history about six centuries into the first millennium of Christianity. For Muslims, the faith they live as Islam has no beginning and no end. From an inside view such impositioned boundaries border on the absurd. Islam's messenger, the Prophet Muhammad, is recognized as the final and last word in a series of prophets through whom the God of Abraham, Moses, Jesus, and Muhammad deals with the humans "He" alone created, breathing spiritual meaning into Mesopotamian clay. Thus—for outsiders to know, even if they cannot accept it as their own inner truth—Muhammad was chosen as messenger of the divine message of the Quran, which is regarded by Muslims as the last literal word in intrinsically untranslatable Arabic from the same God spoken of earlier by Jews and Christians. Whether you are Muslim or not, the subsequent history of much of the known world from our still entrenched Eurocentric perspective revolves in large part around the influence of Islam and the diverse cultures that embraced it, absorbed it, spread it, and still revere it. Just as it was pragmatically prudent a thousand years ago for a Christendom-inated Europe to observe Muslims in the real world, so it is now after another Armageddon-less millennium change. But hopefully we can get beyond intellectual remakes of the Crusades.

For anyone interested in how Muslims have observed themselves, there is a massive corpus of original material, most of which is untranslated for a Western audience. Fortunately Arab, Persian, and Turkish scholars were interested in their own cultures long before Western scholars came along. The sheer bulk of indigenous interpretation of Islam from within demands caution from those who would sum up a widely dispersed historical faith into an essentialized package of ideal beliefs. Some Muslims act as if there is only one true Islam—generally their own variety—not unlike the born-again Baptists who foreswear there is but one true and precipitously narrow way to Christ. A charitable framing of history suggests that there have been many ways of observing and representing what has become one of the world's major religions. Which way is right? Which is the straighter path? If we may borrow the cross-culturally relevant metaphor about the blind Hindu and the elephant, it may depend more on what you feel than what you are able to "see" on the surface.

Within a hundred years after Muhammad died (ca. 632 C.E.), Muslims had taken their faith westward across the Pyrenees into southern France and eastward to the borders of Samarqand. Christian Europe's discursive interlocutors, facing the immediate political reality of Islam's expansion, understandably reacted negatively to an upstart set of invading "infidels." Islam and its prophet were for the most part vilified without even a pretence of objective observation. A crusader apologist, Guibert of Nogent (ca. 1112 C.E.), disparaged Muhammad as so profound a profligate that "it is safe to speak evil of one whose malignity exceeds whatever ill can be spoken."[20] Yet, despite this vitriolic rhetoric *ad extremum*, West and East engaged in extensive economic, technological, cultural, and intellectual borrowing. Monotheistically inclined Christians, Muslims, and Jews had much in common, beyond a mutual dislike. Even some of the quintessential aspects of a nascent European scholasticism and humanism were not immune to Islamic influence. Graduate students at modern-day Harvard University or Hofstra University might be surprised to learn that their "doctorate" comes via medieval Latin (*licentia docendi*) from the classical Arab university term for a "license to teach."[21] The Renaissance discovery, or shall one say invention, of its classical roots stemmed in no small part from Arabic sources, including translations and commentaries of and on increasingly seminal Greek and Latin texts. The eventual economic and overt colonial expansion into the Middle East by European powers, coupled with the rise of an academic structure for interpreting the "Orient," fueled a passion for observing the region that cradled Judaism, Christianity, Islam, and, by extension, "civilization" itself. But, in typically backward historical hindsight most observations recorded till a century ago are rarely up to the objective standards we pose, or at least suppose, a *causa sine qua non* for modern scholarship.

Introductory books about Islam often begin with a ritualized list of the "five pillars," providing an easy fill-in question for standard tests given students in Islam 101 classes. These five pillars, a useful bundling that postdates the time of Muhammad, say little about the message of Islam, apart from the *shahada*, the witness to there being only one God and one final prophet. Prayer (*salat*), alms (*zakat*), fasting (*sawm*), and pilgrimage (*hajj*) are duties, highly symbolic religious acts necessary for Muslim observance. Missing from this picture, however, is an expanded creed, a statement of faith that would flesh out this ritual count. Certainly the *shahada* is the central message of Islam, but only in the boiled-down sense of John 3:16 for born-again Christians. Islam is a monotheism and Muhammad is its definitive prophet: this should be a starting point, not a conclusion. A peculiarly "Western" way of viewing Islam has been to reduce it, following the path of Wilfred Cantwell Smith, to an orthopraxy, a religion united by practice rather than shared belief.[22] Given the extraordinary depth of Islamic thought, is the idea of an Islamic orthodoxy really so toxic to non-Muslims? Muslim theologians formed a complex set of beliefs from the message revealed in the Quran and statements (*hadith* literature) attributed to Muhammad. The Quran speaks of morality, the cosmic battle between good and evil, the wiles of the Devil, the resurrection of the dead, judgment day, the relationship between men and women, and many practical aspects of daily life. To assume that the adumbrated five pillars is analogous to the Ten Commandments or the Nicene Creed is thus to shortchange knowledge of the very beliefs that make the ritual duties significant. What Muslims have done with their *sunna* is certainly as doctrinally relevant as what scholastic icon Saint Thomas Aquinas did with his *Summa*.

If the reader wants to know what Muslims believe, the best way is to ask Muslims themselves.[23] With estimates now rising well over a billion, minus the obvious high percentage of those still being taught how to be Muslim, this is not hard to do. After over fourteen centuries of existence and global expansion there is a diverse and widely variant range of views that have been in one way or another defined as Islamic. Islam has long been an active missionary religion, so there are books and tracts with particular doctrinal and political spins in all the major languages. Some of these have been translated into English and other European languages; several of the basic theological texts are now available on the Internet. It is easy to find information about Islam; the trick is sifting through the rhetoric that represents the religion. Inevitably, pragmatic visions of tolerance aside, the nature of Islam as a revelation claiming to provide ultimate truth for all humanity results in a competition with other universalistic religions, such as Christianity, as well as those particular religious groups who desire only to be left alone and unconverted. My concern, as an anthropologist, is not with entering into this subjective and emotionally charged fray, but simply assessing how those of us who study Muslims in ethnographic context represent "Islam" as such.

Anthropology has already played a role in showing where Islam definitely does *not* come from. Popular stereotype, reinforced by indiscriminate reading of Arab savants like Ibn Khaldun, has long disparaged Islam as a fanatical faith born in the desert.[24] Thus, the ghosts of Bedouins past haunt an allegedly monolithic "Arab mind" through a never-ending cycle of feuds and raids that define the newer monotheism as violent and uncivil. "Desert Ishmael cannot trust," avers one historian of religion in the 1920s, "he cannot co-operate, he cannot cohere in any permanent organization of which mutual faith is the essential cement."[25] The idea of birth in a barren landscape no doubt symbolizes the apologetic thrust of Christians who also found the religious doctrines and rituals barren to the point of heresy. The common misperception of a "nomadic" origin for the religion of ancient Israel has been equally and egregiously applied to the ensuing monotheism of Islam. But Muhammad was no Bedouin; Mecca and Medina were certainly not dung-speckled camel stops in the wilderness. The more ethnography has been conducted among contemporary Bedouin nomads, the clearer it has become that this contemporary tribal society is not very instructive for reconstructing the early Islamic community. As a result, anthropologist Eric Wolf proposed a socially scientific model for the origins of Islam in its urban economic and political aspects; this was half a century ago and might as well be buried in a Near Eastern tell.

Where Anthropologists Come From

Judging from the relatively isolated status of anthropological writing in contemporary academe, it is more profitable to suggest where anthropology appears to be heading than where it has come from. I suppose this is only to be expected from a discipline that informs beginning students—at least in America—that it can and will deal with all aspects of cross-cultural humanity. Anthropologists presume to touch on human evolution, all of human history, ecology, economics, politics, religion, psychology, linguistics, and just about any subject that also has a special discipline devoted to it. To top it off, anthropology departments are often forced to share university space with the seemingly more respectable, at least in a statistical sense, field of sociology. The sibling rivalry between anthropology and sociology, no longer comfortably divided by those who work on one specific side of the civilizational divide, is further muddied by the perennial interest anthropologists have in founding-father sociological icons, such as Durkheim and Weber. This is less of a problem in Europe and Britain, where anthropology was long ago relieved of the burden of being holistic. Indeed, it is a moot point whether British scholar Ernest Gellner is a sociologist who did fieldwork in Morocco or an anthropologist who writes like a hybrid cross between sociologist and social philosopher. The key point is that both anthropologists and sociologists study humans as social or cultural beings; anthropologists always share the stage with other trained scholars.

To shorten a rather long and complicated intellectual genealogy, it is best to remember that in effect anthropology comes from the field. The earliest textual consolidation of the new field, Edward Tylor's *Anthropology* (1881), differed from earlier accounts of customs by stressing the need for observing "primitive" peoples in an objective and scientific way; this was also styled ethnology.[26] The discipline began haphazardly, as "anthropological" texts in the nineteenth century tended to be heavy on speculation and reliant on suspect observations by missionaries and travelers.[27] Modern ethnography—reporting on people whom the anthropologist had lived among—was brought into full, public view with the published work of Bronislaw Malinowski out of his World War I exile on the Trobriand Islands. The fieldwork focus was also fanned by Franz Boas, who trained the first generation of American ethnographers at Columbia University. The earliest fieldwork worthy of the name was conducted among so-called primitive peoples rather than the civilizations of Asia or Europe itself. The ethnographic encounter with Muslims would have to wait until almost the middle of the last century.

The intellectual history of anthropology as a modern discipline is beset with several problems. First, the term itself has been used for philosophical speculation on humans as well as for almost anyone who traveled and wrote about exotic others encountered along the way. Thus, as John Zammito argues, Immanuel Kant and Johann Gottfried Herder can be legitimately claimed as conceptual forebearers for the "calving" of anthropology from philosophy.[28] The Victorian traveler Richard Burton considered his annotations on life he observed in the Middle East, Africa, and India as "anthropological," prompting literary critic Patrick Brantlinger to chide modern anthropologists for not including this Victorian rapscallion in the pantheon.[29] But simple observation does not make untrained dilettantes into methodological role models. Second, the academic arc of a Spencerian—only more recently, Darwinian—"social evolution" dims the positivist glow in the early speculations of scholars like Edward Tylor and Lewis Henry Morgan, who are in the canon of anthropological founders. In attempting to view man as up from the ape rather than a little lower than the angels, there erupted a mutual antipathy between anthropologists and theologians. When ethnographers eventually left their armchairs for the bush, the direct encounter there with missionaries only heightened the chasm. Thus, theologically minded scholars with an interest in Islam were ill disposed in the past to read the pioneering works of anthropologists. A third factor arises with the post-colonial critique of anthropological fieldwork following the break-up of direct European colonization. Far from being seen as an objective recording of exotic cultures for posterity, ethnography as a textual genre of representation has come under fire for fixing a very Western image of the non-Western other. In such a climate, as Edward Said charged little more than a decade ago, anthropology may be damned to get posted always on the wrong side of the "imperial divide."[30] Of course, defining the divide can go both ways.

For the reader interested in knowing what is going on in anthropological theory right now, as I write this text, it is only necessary to look at the critical currents eroding the established moorings of disciplines and fields across the social sciences and humanities. Other scholars may not have been reading many anthropologists apart from Clifford Geertz or a nostalgic reread of Margaret Mead, but many anthropologists have been reading widely from postmodern interpreters: Adorno, Anderson, Bakhtin, Barthes, Bhabha, Derrida, Foucault, Gadamer, Habermas, Jameson, Said, Todorov, and others do appear on reading lists in contemporary anthropological theory.[31] Indeed, at times it seems as though nonanthropologists get precedence over the founding fathers and guiding lights actually trained in the discipline. Much of the current turmoil is an American problem. The halcyon days of a four-field approach that insisted anthropology could be a holistic study of humans have been clouded over by the realities of trying to pull off such a magic trick and at the same time escape the blistering attack on all scientific pretensions to objectivity. Can one discipline train individuals to have skills in linguistics, ethnography, archaeology, and biological evolution? Does specialization not make a mockery of such a goal, especially when other disciplines often do virtually the same kinds of analyses?

Anthropology, like several disciplines, is emerging from a postmodern critique that attempted to virtually do away with it. "Postmodern ethnography advocates the deconstruction of anthropology, especially from without, particularly through literary theory," exclaims Dan Handleman with rightful indignation.[32] The problem in taking apart "anthropology" is figuring out what kind of anthropology is being targeted. Much of the criticism has been needed, but often it has overidentified flaws of individual scholars or focused exclusively on inconsistencies in texts with glaring theoretical faults. My aim here is not to defend anthropology as a discipline, especially American anthropology, or indulge in speculation about what anthropology should be. Those who are not practicing anthropologists should know that reading seminal texts or much-vaunted critiques will give you as much knowledge of what is happening at ground level in the discipline, as reading a theological text will mirror the pragmatic behavior of believers. Rumors to the contrary, neither the scientific investigation of human origins nor the ethnographic observation of human behavior in non-Western societies is about to be aborted.

A central concern of this book is to explain a certain area of anthropological representation to non-anthropologists and anthropological colleagues who are interested in the subject of Islam. It is important to remember that interest in the "ethnology" or customs of contemporary Oriental peoples, especially Muslims, has a long pre-anthropological history. Medieval travelers, including pious pilgrims, crusading knights, and merchants, sometimes left accounts of the Muslims they passed by. Ironically, the most well-traveled medieval representation of Muslims was *The Travels of Sir John Mandeville*, a fictitious autobiographical account attributed to a fourteenth-century English knight. The irony is not that an errant knight would describe Islam, but that he would do so in a rather favorable light, remarking that Muslims were more devout and honest to their religion than Christians of his day had become to their faith.[33] By the nineteenth century Christian missionaries had settled into the Holy Land and produced hundreds of books with titles like William Thomson's (1858) *The Land and the Book or, Biblical Illustrations Drawn from the Manners and Customs, the Scenes and Scenery of the Holy Land*. The thrust of most of this literature was unabashedly apologetic. "The remarkable reproduction of Biblical life in the East of our day is an unanswerable argument for the authority of the sacred writings" avers Rev. Henry J. Van-Lennep; "they could not have been written in any other country, not by any other people than Orientals."[34] The Bedouin sheikh, feared as a thief while actually traveling in Palestine and Syria, still served as a potent reminder of how the biblical patriarchs must have lived. Nomads, camels, tents, veiled wives: the stereotyping of Muslims proceeded textually by imagining an idyll of those who were probably least devout in the region. There are also the custom-packed accounts of swashbuckling adventurers, most notably Richard Burton's description of a surreptitious trip to Mecca in 1853, replete with maps and illustrations exposing the most sacred site of Muslims. So extensive was this corpus of traveling texts that the bibliography of a major anthology on *Peoples and Cultures of the Middle East: An Anthropological Reader*, published more than a century after Burton's text, contains more references to travel accounts than to ethnographies.[35]

Until the 1970s there was little anthropological discussion of "Islam" as a religion in Middle East ethnography. The Dutch scholar Christian Snouck Hurgronje and the Finnish sociologist Edward Westermarck were among several individuals who wrote about Islam from firsthand experience, but not in the modern sense of participant observation for an extended period of time *in situ*.[36] The first modern ethnographic account of an Islamic context may be Evans-Pritchard's *The Sanusi of Cyrenaica*, based on field research in Cyrenaica during World War II.[37] Eric Wolf, a cultural anthropologist, published his "functional" argument in 1951 about the origins of Islam in Mecca, but he had never conducted research in an Islamic context, nor did he know the plentiful Arabic sources firsthand.[38] This thesis was followed up by Barbara Aswad, who cited Orientalist sources rather than drawing on her fieldwork in a Syrian village.[39] The respective authors of a 1955 article entitled "Zur Anthropologie des Islam" and a 1961 article labeled "An Analysis of Islamic Civilization and Cultural Anthropology" turn out to be two Arabist historians, Annemarie Schimmel and Gustave von Grunebaum. The French scholar Joseph Chelhod carried out limited ethnographic research in the 1960s, but he was primarily an Arabist teasing anthropological insights out of Arabic texts.[40]

Why was there so little anthropological interest in Islam? Michael Gilsenen, reflecting on his anthropological education at Oxford in the 1960s, asks "Where was the 'Middle East'?"; his blunt but astute answer is "nowhere."[41] At the time academic rendering of Islam was still in the hands of Orientalists, scholars who could read the literature in Arabic or the relevant language, historians, and the evolving field of religious studies. Up to that point, probably the most widely distributed anthropological book on the Middle East as a whole had been Carlton Coon's *Caravan* (1951); Coon was in fact a physical anthropologist who visited Yemen briefly but never conducted ethnographic fieldwork.[42] The start of the 1960s witnessed the first installment of the revised *Encyclopaedia of Islam*, the most authoritative—in the strict sense—representation of Islam in a Western language. Yet virtually none of the articles in the first new volume were written by ethnographers.[43] Richard Antoun, in a mid-1970s survey of prior anthropological research, explained the reluctance of anthropologists to deal with Islam as a result of disciplinal tunnel vision—leaving Islam to the Orientalists—and the poor Arabic language skills of fieldworkers.[44] At the same time, Muslim anthropologist Abdul Hamid el-Zein provided a critique of previous anthropological discussion of Islam, making a provocative case for a refined study of the various "islams" actually observable.[45]

In historiographic hindsight, ethnographic work among Muslims across North Africa, the Middle East, and Central Asia can be divided by the evolving paradigms within the discipline as a whole.[46] The early bias for working among so-called primitive peoples as a kind of living laboratory for reconstructing human history insured that an ethnographer was more likely to seek out a New Guinea tribe than a village of Muslim peasants. Much of the published work is primarily concerned with social and cultural issues with only occasional emphasis on Islamic doctrine or practice. Those anthropologists who did choose the Middle East or North Africa focused on tribal nomads at first, influenced in large part by the seminal kinship studies of Evans-Pritchard on the Sudanese Nuer and Cyrenaican Bedouin. In addition to work on North Africa, by the 1960s ethnographic studies of pastoral Muslims began to appear for Iran, Pakistan, Arabia, Turkey, and Israel.[47] The tribal link of Muslims, so long the stereotype outside anthropology, fit nicely with the compulsive interest early anthropologists had for kinship.

At the same time ethnographers also began to indulge in descriptive community studies, influenced by Robert Redfield and a penchant in Latin American ethnography to look at peasants. Here the Western observer rented a room in the village and in good faith tried to record virtually everything going on. Apart from a few studies by Arabic-speaking anthropologists like Hamed Ammar and Richard Antoun, most of the earlier community ethnographies bear witness to the researcher's meager language skills; representation of Islam was almost always brief.[48] A representative example of this inattention can be found in Louise Sweet's ethnography of a Syrian village. In a text of about 250 pages, only about 4 pages are devoted to Islam.[49] The very terse description is mainly about local practice of the "five pillars of Islam," with a conclusion that the "condition of supernatural beliefs and religious organization" in this Syrian village "seems weak" in comparison with the other description available—but not examined—from earlier French writers in the region. The impression left is that these hard-working peasants had little interest in being pious. In a similar way, Paul Stirling's study of a Turkish village treats Islam as though it had been superceded by the rhetoric of nationalism. Quranic teaching is characterized as incomprehensible and easily replaced by the modern government.[50] The entries under "Islam" in his ethnography's index are that it was "disestablished," followed by "treachery of" and "victory of over Greeks." Occasionally an entire culture area was examined, such as Algeria, Afghanistan, and the Hindu Kush, although here again Islam generally fades into the cultural background.[51]

If there is one dominant theme connecting ethnography with Islam, primarily in the North African context, this would be Islamic mystics, Sufis, and marabouts, starting with Evans-Pritchard's pioneering study of the Sanusiyya order in Libya. Another British anthropologist, Ernest Gellner, wrote a major study of the political role of marabouts, playing off the more orthodox "doctor" of great-tradition Islamic law against the little-tradition rural "saint."[52] Clifford Geertz looked at sufism as a symbolic dimension of Islam in Morocco and Indonesia in his seminal *Islam Observed*. One of his students, Vincent Crapanzano, condensed the discussion of a mystic named Tuhami to a psychoanalytic portrait,[53] while another student, Dale Eickelman, detailed life in a Moroccan pilgrimage center.[54]

On the other English-speaking side of the Atlantic, I. M. Lewis surveyed sufism in Somalia and Michael Gilsenen analysed an Egyptian
sufi order in Egypt.[55] Once again, it seems as though anthropology
came to Islam via the exotic, as though the mundane was too
obvious, perhaps too boring, to require explanation.

Without question, the most comprehensive study of Islam in a
local community from this period is Egyptian anthropologist el-Zein's
ethnography on the religious aristocracy of Lamu, an island off
Kenya.[56] The value of el-Zein's analysis is that the religious symbols
and retold myths he encountered are contextualized within the defin-
able social structure and observable behavior of real Muslims. Moving
beyond the Durkheimian ritual of searching for the social function in
local religion, el-Zein probes the pragmatic hermeneutics of masters
and slaves in ritualizing Islamic stories of creation and the prophet
Muhammad. The ethnographer provides a painstakingly detailed
account of how Islam is practiced. An entire chapter is devoted to his
observation and participation in Ramadan mosque readings of the
Swahili creation myth.[57] We learn not only what various local groups
say the myth means to them, but the seating arrangements of those
who attend, their greeting behavior on the way to and from the
mosque, even the role of children. Text is wedded to context in a
manner no study of Islam had previously achieved.

The last three decades of twentieth century ethnographic studies
reveal a major shift in approaching Islamic cultures. Paralleling a
trend in the discipline at large, female ethnographers began publishing on issues of gender and sexuality. These included anthropologists
with their own genealogical roots to Muslim societies, such as Lila
Abu-Lughod, Soraya Altorki, Fadwa El Guindi, Shahla Haeri, Ziba
Mir-Hosseini, and Fatima Mernissi, to name but a few. Impressive
contextualization of Islamic ritual in African societies can be found in
the work of Ladislav Holy, Michael Lambek, and Robert Launay, as
well as in Indonesia, exemplified in the studies of John Bowen, Robert
Hefner, and Mark Woodward. There are now specific studies of the
major rituals, such as pilgrimage, fasting, prayer, *mawlid* (celebration
of the Prophet's birthday) and sacrifice.[58] Egyptian anthropologist
el-Sayyed el-Aswad offers a detailed ethnographic portrait of folk
cosmology in an Egyptian village.[59] Some anthropologists have also
linked their fieldwork to the literate tradition, such as Brinkley
Messick's study of legal texts and decision making in the Yemeni town
of Ibb or my own work on Islamic folk astronomy and agriculture.[60]
In more recent years anthropological studies have appeared on the
resurgence of a politicized Islam, starting with the aftermath of
the Iranian revolution and expanding to the global phenomenon of

"Islamism" now dominating the news.[61] Even Salman Rushdie's *The Satanic Verses* gets anthropological coverage.[62]

Contemporary ethnographers are attuned to the global dimensions of popular culture in Islamic societies. In a volume on *New Media in the Muslim World: the Emerging Public Sphere*, Dale Eickelman and Jon Anderson suggest that "by looking at the intricate multiplicity of horizontal relationships, especially among the rapidly increasing numbers of beneficiaries of mass education, new messages, and new communication media, one discovers alternative ways of thinking about Islam, acting on Islamic principles, and creating sense of community and public space."[63] John Bowen examines contemporary religious poetry in Sumatra.[64] Presentation of Islamic issues in the print media has been one of the major concerns in the work of David Edwards in Afghanistan and Gregory Starrett in Egypt.[65] Anthropologists are also participants observing the same television programs, films, and videos as their informants.[66] Jon Anderson, whose original fieldwork was in Afghanistan, has traced the trajectory of Islamic websites on the Internet from creolized pioneers, especially Muslim graduate students, to more officializing discourses of formal institutions.[67] Ethnography has come a long way from its kinship tunnel vision with Bedouins on camelback.

Where the Author Comes From

Significantly, the texts I have selected for analysis in this study are those that were influential in one way or another in my own training as an anthropologist preparing to do ethnography among Yemeni Muslims and my subsequent research trajectory in Middle East anthropology, the history of Islam, and editing "medieval"[68] Islamic texts. The introduction to observing "Islam" through my discipline's lens was *Islam Observed* by Clifford Geertz, who had established his reputation as an essayist drawing from ethnographic fieldwork in Indonesia and Morocco, two Islamic countries at opposite ends of a geographic spectrum. True to his anthropological roots, although relying heavily on a particular reading of hermeneutics, Geertz acknowledges that what he saw in "the broad sweep of social history" he first saw or thought he saw in "the narrow confines of country towns and peasant villages."[69] Ernest Gellner, another anthropologist whose work I examine, once remarked that "Anthropologists are at home in villages" just as "Orientalists are at home with texts."[70] In an academic career that commenced the same year as Gellner published this comment, I have tried to feel at home in both.

With all due respect to Geertz and Gellner, I encountered Islam long before I met Muslims. At first, it was the childhood attraction to the Islam of the Arabian Nights, Sir Richard Burton's Meccan escapades, and my grandmother's attic full of *National Geographic*. Then it was the Islam of Orientalist scholars, as I applied myself to learning classical Arabic as a graduate student. Marrying the granddaughter of a Syrian qadi made more real the Islam I had been absorbing piecemeal. Yet, when I arrived in Yemen in 1978 to begin eighteen months of ethnographic fieldwork on traditional irrigation and water rights, my understanding of Islam was still obscured by biased misrepresentations, well meaning but partial representations and a simple lack of exposure. In the field I learned quite quickly that what Muslims do is not reducible to what books, even their own, say they should do.[71] But I also developed a passion for knowing what the texts were saying. As a personal confession, I have at times felt more at home in manuscript libraries poring over esoteric Arabic texts than playing the obvious outsider talking with villagers. Working with texts, I have developed a healthy—at least in my mind—skepticism about the "truth" of anything written. Working with people who think they have the truth only makes it more evident that texts and words should be approached as means and not ends.

The Islam I observed in a highland Yemeni village was not just the routinized five pillar variety. All of the major rituals involved were obviously part of the context, but the value of living there was that I could observe the day-by-day behavior of real people rather than just read about the ideals. As an outsider, an American at that, I did not feel out of place or on display. Perhaps I was fortunate to live in an isolated village before the political madness of suicide bombings. The fact that my wife, also an anthropologist, is a native Arab speaker and Muslim certainly helped smooth our joint acceptance as resident outsiders. Maybe it was also the mundane focus of my research on agriculture and irrigation, readily understandable on a pragmatic level by these expert farmers. I had not come to Yemen to study "Islam," but neither had I decided ahead of time to ignore anything that came along in the process of being there. The men I spent the bulk of my time with were rarely involved in theological disputation, certainly not with me. What they did or chose not to do was their Islam and that, at least at the time, seemed the most natural thing in the world.

If I were only a historian, it would be tempting to regard individual Muslims primarily as useful informants for interpreting difficult texts. I feel quite different after living with Muslims in several Arab countries. As much as I love history, the anthropologist in me reads others first as fellow humans sharing an ancestry to a distant past long before any of the formal religions of today were formed. Thus, to be a Muslim is to be so in a specific time and place, to live like anyone else through the cycle from birth to death. In this sense the notion of "observing" Islam is a telling oxymoron, especially from the angle of an anthropologist who has been to the field. Muslims can be observed; their material culture can be documented; their words read and lexicated; their behavior witnessed. "Islam" in the abstract sense of a religion or civilization can only be represented. The temptation to converge thought and sight is so engrained in our linguistic usage that representation easily gets reified as what must really be there. "Being there," for me, broke the spell of this epistemological chimera. There are good Muslims and bad Muslims, devout Muslims and indifferent Muslims. These do not get properly represented under the umbrella of an essentialized and homogenized "Islam." If the only Islam wanted is the ideal, then read a book or listen to a sermon. For the common humanity shared by us all, regardless of religion, culture, or ethnicity, it is necessary to observe others as they live their lives or read the arguments of those who have done exactly this.

This book, oddly enough for an otherwise old-schooled anthropologist, is a text about texts. Unlike most of the anthropologists mentioned above, I do not describe my own fieldwork in any depth. Nor do I sum up in a tidy fashion the accumulated knowledge that could be fit into an authoritative anthropology of Islam. I assume here the role of a literary critic informed by my experience as a participant observer of Muslims and an avid reader of ethnographic texts. In looking over the sum of the following chapters, I am struck by the amount of criticism leveled, a little like a student's initial panic at the anticipated flow of the professor's red ink over a carefully typed essay. It is easy to attack, to flail monologue style at absent authors; this is the ease by which much postmodern critique has evolved Borg-like over the past several decades. Who, after all, do I think I am, bashing arguments made by such renowned scholars in my own field as Clifford Geertz, Ernest Gellner, Fatima Mernissi, and Akbar Ahmed? My own rhetorical iconoclasm could easily fixate on the bearers of the messages, *ad hominizing* them, rather than challenging their ideas with the same spirited engagement these authors criticized those who came before them. My intention is otherwise, to lay bare rhetorical veins rather than add to the current print terrorism against the vibrant faith of Muslims. Islam will be represented and this is one attempt to look at how four anthropological texts do just that, but obviously far more than that.

Chapter 1

Clifford Geertz: *Islam Observed* Again

But can one ever imagine Islam without Muslims?[1]
Ebrahim Moosa

The dynamic global growth of Islam as a major world religion shows few signs of abeyance. Muslims have been around for the better part of fourteen centuries, although trained ethnographers have been observing them for less than a century. To the extent anthropologists are interested in the globalization of religious communities, as well as in those small and more manageable locales where ethnography must be done, it is not surprising that the academic moniker of an "anthropology of Islam" has evolved. There is, of course, an important difference between studying people who happen to be Muslim and consciously formulating arguments about Islam as a religious system. The latter has been dominated by historians and philologists who study Islamic texts. As Edward Said's brash branding of "Orientalism" indicates, many of the Western scholars who studied Islam brought with them the baggage of ethnocentric and racist biases. But Said, who doles out praise rather rarely, also suggests that there have been a few exceptions, notably in "the anthropology of Clifford Geertz, whose interest in Islam is discrete and concrete enough to be animated by the specific societies and problems he studies and not by the rituals, preconceptions, and doctrines of Orientalism."[2] If at least one anthropologist could escape the manacles of Orientalist discourse, the idea of an anthropology of Islam is certainly worth exploring.

It has become commonplace without and within the discipline of anthropology to locate the jump-start of a specific "anthropology" of Islam with Geertz's well-known and widely read *Islam Observed*. In an early survey of the topic, Abdel Hamid el-Zein chooses Geertz's text as a starting point because of its "scope and sophistication."[3]

Robert Launay cites this text as heralding the "birth" of an anthropology of Islam.[4] Although critical, Henry Munson labels it "one of the earliest attempts by an American anthropologist to bridge the gap between the *histoire des mentalités* and anthropology."[5] Such praise, certainly not undeserved, parallels the canonization of Geertz in the advanced institutional studies sense, as "the most influential anthropologist of his generation" and the super[ceding]star of Margaret Mead as the intellectual ambassador-at-large from anthropology.[6] Even the casual browser of a corner Barnes and Noble bookstore can attest the readability of the Geertzian corpus to the world at large.

Following on an earlier essay about religion as a cultural system, Geertz evolved into the paradigm of an ethnographer who goes beyond local description of Muslims to a thicker—and definitely slicker—"interpretive" or hermeneutic approach to "Islam" as such. Ironically, if this is indeed where anthropologists want to situate their theoretical interest in Islam, Geertz derives his theoretical mooring and antitheoretical musing more from sociological and philosophical than anthropological roots. As Geertz freely admits at the start of the lectures that comprise *Islam Observed*, his key intellectual ancestors include Max Weber, through the media of sociologists like Talcott Parsons, Edward Shils, and Robert Bellah, and historian of religion Wilfred Cantwell Smith. Indeed, he asks for his ideas to be judged not on their ethnographic merit—how accurately they reflect the lived reality of Muslims—but as part of the program of "comparative, historical, macro-sociology."[7] This program, along with the aura of authoritative expertise that pushed for it, faded from view in the poststructuralist horizon dawning as *Islam Obscured* was published. With the metabolic arrest of meta-theorizing, Geertz's call for a symbol-driven reading of Islam-as-a-cultural-system has reflexively been disenfranchised. At least this has been the case in Geertz's native discipline of anthropology, where the mantle of ethnographic authority—worn out in print but still worn with pride by many in the field—has devolved into the mandala of the Emperor "Being There's" New Clothes.

If there is to be a consensual anthropology of Islam that involves observing Muslims, regardless of the discipline of the observer, then *Islam Observed* needs to be observed yet again. "Again" is the key word here, because this seminal text has survived the thick and thin of criticism from the laudatory to the disclamatory. "In *Islam Observed*," notes Richard Martin, "he [Geertz] stepped beyond the bounds of village anthropology to consider the more expansive notions of Islam held by orientalists and historians of religion, offering constructive criticism of venerable theories and categories in

Islamic studies."[8] The village anthropologist, previously regarded as a village idiot by many humanists, stopped collecting data up in the trees long enough to go up the mountain and talk about the meaning of the forest. In this case an anthropologist carried a repertoire of hermeneutic tools, a desire to see Islam beyond the comfortable but constrictive borders of the informant's observable world. It is hardly a surprise, then, that Geertzian anthropology steadily gained converts outside his discipline, even as ethnographers continued to observe other Muslims and interpret localized versions of Islam in different ways. I have one overriding rhetorical query to add to the discussion: in leaving the village for the pundit's authoritative and lecture[ous] perch, where did all the ethnography go?

Observations on *Islam Observed*

I have attempted both to lay out a general framework for the comparative analysis of religion and to apply it to a study of the development of a supposedly single creed, Islam, in two quite contrasting civilizations, the Indonesian and the Moroccan.[9]

Clifford Geertz

In 1967 Clifford Geertz was invited to speak in the Terry Foundation Lectures on Religion and Science at Yale University. Considering how ephemeral academic lectures are and how stale lecture series—even at Yale—can become, it is remarkable how successful Geertz's published version remains. There is probably no other book by an anthropologist on Islam that has been read by a wider circle of students and scholars within and without the discipline of anthropology. Raymond Firth's comments, quoted on the back cover of my original paperback edition, now seem prophetic in the oh-so-quotable prediction that "it should keep students of Asian societies and of comparative religion busy for a long time." Extensive commentary and the sheer number of citations stretching over three decades since publication abundantly prove this to be the case. It is worth asking, given its humble academic origins, why this particular book has enjoyed such a long and prosperous shelf life. Is it the brilliance of the argument, the eloquence of the prose, the truth of the claims or the unshakeable stature of the author?

For those who have not read *Islam Observed* recently, a short summary of the book's goals and plan could be useful. In about one hundred pages, Geertz lectures on the theme of the book's subtitle: "Religious Development in Morocco and Indonesia." The goal of the study is twofold, both to elaborate his "general framework for the comparative analysis of religion" and to illustrate this through Islam "in two quite contrasting civilizations."[10] For Geertz religion is sustained by its "symbolic forms and social arrangements." His first chapter contrasts, in the full sense of the word, tribal Morocco with peasant Java in Indonesia. A second chapter sums up what Geertz sees as the "classical religious styles" of Moroccan and Indonesian Islam, in large part by striking off, as he puts it, an Indonesian apostle prince and a Moroccan saint. This classical stasis is interrupted by a "scripturalist interlude" of three processes acting in both geographical areas. First, the West comes to dominate the cultural context of Islam, second, there is an increasing local reliance on the "scholastic, legalistic, and doctrinal" parts of Islamic tradition, and finally the modern nation-state emerges.[11] The final chapter concludes with Geertz's customary common-sense approach to religion as a "struggle for the real" in which the reality for Muslims centers on change occuring in a political world where King Muhammad V serves as a counter symbol to President Sukarno. In the end, at least of the lectures, "amid great changes, great dilemmas persist." Moroccan Islam has advanced "almost to the point of spiritual schizophrenia" and Indonesian Islam has been absorbed into "a cloud of illusive symbols and vacuous abstractions."[12] This is the gist, minus the historical detail and rhetorical flair.

Geertz's observations extend well beyond what he may have observed in Java and Morocco. In fairness, he gave these lectures as an advocate for his discipline. His audience of historians of religion, steeped in texts, was assumed to be suspicious of the odd scholar who had the audacity to expound on Islam from the limited experience of a few modern Muslims in some exotic topos. "We are all special scientists now," Geertz reminded the old-school skeptics, "and our worth, at least in this regard, consists of what we are able to contribute to a task, the understanding of human social life, which no one of us is competent to tackle unassisted."[13] What the lectures try to present is a set of observations based on, one might think, the author's unique experience as anthropologist. Yet, the bulk of his observations are mostly about historical matters and taken from the secondary texts listed in his bibliographic note.[14]

With apologies to Roland Barthes and Michel Foucault, it is still useful to deal intentionally with the author. Geertz concedes that his inspiration for the lectures and subsequent book stemmed from his fieldwork experience with Muslims. In the 1950s Geertz spent about three years in Indonesia, later writing a major ethnographic monograph on the religion of Java.[15] Mark Woodward contends that Geertz's Javan ethnography "is best understood as an elegant restatement and theoretical reformulation of colonial depictions of Islam."[16] In addition he had already begun less extensive fieldwork in Morocco, although not specifically on religion.[17] Geertz parades his own ethnographic authority—the rarity of fieldwork in two distinct Islamic societies—as his theoretical starting point, but "theoretical" it remains throughout the lectures. Fieldwork, Geertz asserts, was for him "intellectually (and not just intellectually) formative, the source not just of discrete hypotheses but of whole patterns of social and cultural interpretation."[18] His overview of social history is said to flow from experience among Muslims in villages and towns. But surely the order here should be reversed, even if accepted as rhetorical flair. Geertz knew Weber's spin on Islam before he ever sat down with a Muslim as informant. He found Ricouer before he stumbled upon Ibn Khaldun. As a broadly trained scholar in Western intellectual history, he entered the field not with a *tabula rasa* to be inscribed by the locals but with a proudly acknowledged set of Ivy [be] League [red] interpretive baggage. This is not to belittle his ethnographic skills, nor his wide-ranging synthetic vision as a social scientist, but simply a reflexivist reality check that applies to all ethnographers. We enter the field with conscious and unconscious models "for," "of," and even "against." We ask the questions, record the notes, and write ethnographic texts in established and often programmatic formats. Our interpretative frame—like a genetic code—is borne with us, not born in the field. Notwithstanding the impact fieldwork can and should have on how we go about interpreting ourselves as well as the omnipresent others we claim at times confidently to know better than they know themselves, ethnographers can never write themselves out of their texts.

Such a truism was not so obvious in the late 1960s when the lectures were presented. The reflexivist critique of ethnography had not yet surfaced, although the ingredients that occasioned it were in full force, notably in Geertz's own prolific writings.[19] A decade later Paul Rabinow laments the fact that fieldwork was one of the great "clan secrets"[20] in his intellectual training. His mentors, including Clifford Geertz, told him it was the "alchemy of fieldwork" that defined the real anthropologist from armchair dilettantes.[21] Eventually, Geertz's own ethnographic authority was challenged in print. Vincent Crapanzano, another fieldworker in Morocco, exposes the thinness of Geertz's ethnographic data in a blistering, no-holds-barred critique of one of Geertz's most classic articles, the thick and somewhat playful description of a Balinese cockfight.[22] In a riveting account commonly consumed by introductory anthropology students, Geertz constructs a composite cockfight and the moods and

motivations he finds in Balinese male love of their cocks. The pun
is fully intended. The ethnographer does not relate actual cockfights
he observed or conversations with specific informants, but rather
pretends to look over the shoulders, or perhaps down the pants, of
the Balinese for the native point of view. "Who told Geertz?"
demands Crapanzano after reading Geertz's construction of what the
cockfight "means" to the Balinese. Turning Geertz's own colorful
phrasing back on him, Crapanzano asserts that there is in fact "no
understanding of the native from the native's point of view."[23] The
problem, by no means unique to Geertz, is that the fact of "being
there" as an anthropologist clearly established his authority, especially
for the audience of scholars of religion at Yale. For most anthropologists in the last two decades this appeal to authority has increasingly
worn thin.

It appears that Clifford Geertz was oblivious to the emergent
challenge to his "ethnographic authority" while he himself was
encouraging the reflective process that led to it.[24] Doubly and even
deep-surfacedly ironic, Geertz deliberately sought to defend his
ethnographic experience against arguments from outside the discipline that ethnography was too parochial and exotic to yield valid
social analysis beyond the local level. "The fact that the anthropologist's insights, such as they are, grow (in part) out of his intensive
fieldwork in particular settings does not, then, in itself invalidate
them,"[25] Geertz passionately argued in *Islam Observed*. He expected
those not baptized by fieldwork experience to question his credentials, but perhaps not criticism from his students and more or less
disciplined colleagues. But he also, at least in a rhetorical sense,
recognized that ethnographic experience alone is not enough. The
key point, *a propos* for the theme of the lecture series, is that "Like all
scientific propositions, anthropological interpretations must be tested
against the material they are designed to interpret . . ."[26] His "empirical conclusions" and "theoretical premises" should be validated "on
how effective they are in so making sense of data from which they
they were neither derived nor for which they were originally
designed."[27] Like many of Geertz's acroterial aphorisms, this comes
across as common scholarly sense. Nietzschean hubris aside, the
default epistemological claim for any modern scholar must be that his
or her ideas make some kind of sense of a shared reality.

One of the recognized accomplishments of Clifford Geertz is the
eloquence and ingenuity in his writing. As Margaret Mead became a
household name in the public media, so Geertz has long been
Ganymede, rather than Crapanzano's Hermes, to English-speaking
literary intelligentsia. His essays routinely appear in major newspapers

and magazines; his essay collections are readily available in local bookstores. Indeed, he may be better read outside his discipline than within it, certainly in recent years.[28] The key to Geertz's rhetorical persuasiveness was noted a long time ago by Harry Benda, who observed that *The Religion of Java* brings Islam to life "in a manner unequaled by a good many dry-as-dust treatises."[29] Geertzian word-play even informs the inspiration for a field of "cultural poetics."[30] His seminal book, *The Interpretation of Cultures*, is a set of independent essays in a "highly-wrought literary style," the format of choice in his recent books.[31] In short, Clifford Geertz is widely read because Geertzian prose is eminently readable. Can we blame Geertz for being an effective and persuasive writer? I should hope not. The problem occurs with the large number of scholars who have uncritically and precariously placed his provocative but at times problematic texts on a paragonal pedestal.

The rhetorical style of *Islam Observed*, although far removed from the cocky sexual metaphors of his "Deep Play" article, is both charming and disarming. Geertz has his way with words, especially seeming and unseemly paradoxes. "Our problem, and it grows worse by the day, is not to define religion, but to find it," challenges Geertz at the start of his first lecture.[32] At a time when even *Time Magazine* wondered if "God is dead" and Harvey Cox's *The Secular City* had supplanted St. Augustine's *Confessions* in most divinity schools, this was an apt adage. Or, just a few paragraphs later, "Religion may be a stone thrown into the world; but it must be a palpable stone and someone must throw it." These are aphorisms so *Readers Digest*-ible and common-sensical that it is hard, at least for me, to read over them without underlining and marking their pres[ci]ence in the margin. To put it another way, these are lectures I would not want to or be able to sleep through.

By calling Geertz's style disarming, I do not automatically impugn his accuracy, nor do I wish to question, *á la* Vincent Crapanzano and James Clifford, his right to write his own way.[33] I do find it ironic that an author so focused on "meaning" should have such an exceptional ability to disguise and prevaricate the meaning of his own statements. A key aspect in the appreciation of Geertz's writing by such a large audience is that it is often possible to read one's own meaning into what he appears to be saying. An example, taken more or less at random, is the following pronouncement on how to study religion:

> But the aim of the systematic study of religion is, or anyway ought to be, not just to describe ideas, acts, and institutions, but to determine just how and in what way particular ideas, acts, and institutions sustain,

fail to sustain, or even inhibit religious faith—that is to say, steadfast attachment to some transtemporal conception of reality.[34]

If you attempt an exegesis of this statement, treating it as gospel rather than just a good schmooz, it would be difficult to pinpoint a specific point of view here. There is the assumption that religion can be and ought to be studied systematically, but what scholarly discipline at the time would not have deemed its own methods as systematic? Of course, who can fault the goal of going beyond mere description, avoiding another catalogue of useless facts, anal-analytical stamp collecting and the like? The issue not addressed is how it would ever be possible only to describe, and not at the same time impose, meaning. The put-down placebo of mere description, a straw argument to be sure, avoids the sticky problem of whether some description is a better fit, perhaps even more scientific, than others. Would it be advisable to avoid description altogether?

The aim proposed by Geertz is to "determine" if what people say and do sustains, fails to sustain, or inhibits "religious faith." This is surely a case for cause-and-effect reasoning, doing what real scientists do in their laboratories. But what does "sustain" or "inhibit" mean outside the shallow Petri dish when applied to a topic as deep and thick as religious behavior? Is this a measure of survival or a scale of relevance? For example, does the religiously sanctioned practice among Muslims of allowing up to four wives sustain or inhibit Islam? And how will the reader fill in the rather large blank of defining "faith"? It sounds like common sense and good scholarship, but the words can be, and are, easily interpretable in ambiguous ways.

One of the rhetorical skills that Geertz has mastered well is the raising and consequent gentle lowering down of paradox gridlock. This is a variant of an old scholastic mode of disputation, akin in more modern terms to what Jean Piaget dubbed "disequilibration." Geertz introduces *Islam Observed* by saying that what he proposes to do is obviously absurd, the idea both "to lay out a general framework for the comparative analysis of religion and to apply it to a study of the development of a supposedly single creed, Islam, in two quite contrasting civilizations, the Indonesian and the Moroccan."[35] Although unstated, the underlying metaphorical paradox is how to see both the forest and the trees at the same time. Geertz acknowledges the problem: "What results can only be too abbreviated to be balanced and too speculative to be demonstrable." He invites the reader to wonder with him how two cultures over two thousand years old can be "compressed into forty thousand words." "And well he should!" any reasonable reader should be thinking. Such a goal is formidable, if not admittedly foolhardy. It is also, as Geertz surely knows, enticing. His cavorting with caveat begs the questioning reader for a chance to work some rhetorical alchemy.

"Yet," Geertz adds in overturning paradox redux, "there is something to be said for sketches as for oils and at the present stage of scholarship on Indonesian and Moroccan Islam (to say nothing of comparative religion, which as a scientific discipline hardly more than merely exists), sketches may be all that can be expected."[36] What Geertz is about to do cannot really be done, but since so little has been done and so little is known, he is going to give it the old college try. He is an ordinary man with a sketch pad, not Michaelangelo. But since the Sistine Chapel is not on the tour, a quickly drawn sketch is assumed to suffice. Now there is nothing inherently wrong with sketching or speculation, exploring new ground or applying ideas from one discipline to another. Science would still be mired in musty museum collections if scientists were as reluctant to speculate to their peers as Darwin was to commit his revolutionary evolutionary ideas to print after his revelation on *The Beagle*. But Geertz offers more than a sketch, whether he intended to or not. This was no ritual sand painting to be appreciated only in the moment of its creativity; this is the text eventually instilled as the foundation piece for a gallery on the anthropology of Islam.

I certainly do not blame Geertz for orienting his lectures the way he did. The very idea of a lecture series on science and religion is in itself propaedeutically paradoxical, if not oxymoronic.[37] Geertz presents what he hopes can be accepted as a step in the right direction toward a "scientific" approach to religion. But, unfortunately, he is poorly poised to do such a "scientific" undertaking.[38] Being aware that two well-documented traditions with long histories cannot be easily condensed into a series of short lectures did not prevent Geertz from going ahead and trying to do it anyway. Despite the claims, *Islam Observed* is neither scientific nor ethnographic. First, there is virtually no analysis of primary texts—certainly not in Arabic—here, no novel historical interpretations based on a thorough survey of all relevant sources, no contextual depth and little cultural thickness of any kind.[39] Second, the ethnographic data appear to have been left back in the village. What we get is Geertz's read; the only natives in sight are those viewed generically through the lens of the absent ethnographer's own highly crafted rhetoric. Flesh-and-blood Muslims are obscured, visible only through cleverly contrived representation and essentialized types.

Following Max Weber, his self-acknowledged intellectual mentor, Geertz idealizes what he is in fact not an expert on. *Islam Observed* is

not about observing Muslims—the mantle of authority for the
author—but musing mentalistically through historical and theological
references in which other kinds of scholars are clearly far more fluent.
Historians of Islam should question Geertz's authority as an anthropologist talking to everyday Muslims and then assuming he has *ipso
facto* become a scholar of Islam, colonialism or modern political
history as such. It is all well and good to speak of writing "a social
history of the imagination,"[40] but why should the merely imaginable
be passed off as what might best fit the case? For example, there is a
general consensus among anthropologists who have recently studied
Indonesian Islam that Geertz was overly influenced by a modernist
view that most Muslims in Java were nominal because they did not fit
an idealized sense of orthodoxy.[41] Similarly, much of the analysis of
Moroccan Islam is derivative and reliant on unsympathetic, not yet
de-Orientalized, French scholars.

Ultimately it is Weber's model of ideal types that compromises
Geertz's understanding of Islam in relation to his experience with
Muslims through ethnographic encounter.[42] Geertz constructs several conceptual themes for distinguishing Islam between Indonesia
and Morocco. Islam, in the broadest sense, can be "maraboutic" or
"illuminist"—referring to the mystical bent of North African
sufism—or "scripturalist" in reference to the syncretist intellectualism
of Indonesia. Geertz implies that this is "the" critical difference,
perhaps due in part to the severity of the contrast. In an almost ludi-
crously Lévi-Straussian trope, Indonesia and Morocco are overideal-
ized through myriad contrasting pairs. Consider the prominent
pairing of assumed opposites presented in the first lecture:[43]

Indonesia	**Morocco**
peasant	tribal
wet rice cultivation	dry farming
inward and docile farmers	aggressive sheikhs
built on diligence	built on nerve
Islam adopted civilization	Islam constructed civilization
Islam came by trade	Islam came by conquest
cultural diversification	cultural homogenization
syncretistic	uncompromising rigorism
comprehensive	pure
largeness of spirit	intensity
intellectualism	formalism
reflective	rigorous
multifarious	dogmatic

While similarities are also mentioned, it is the unmitigated, bi-polar opposition that streams through this narrative flow. These constructs are not supported in his lectures through documentation; it is apparently enough that Geertz as expert has taken the time to distill what is relevant from the bundle of references appended as generic bibliographical notes. Geertz attempts to dispel criticism by declaring he has made no attempt to make his "arguments look less controversial, speculative, or inferential than they are by appending to them an extensive list of arcane references . . ."[44] His bibliographic notes are not designed to "support" his interpretations but as an aid for anyone who might be interested in pursuing further study. We should not fault Geertz for being overtly dishonest, since over and over again he makes no claims for being an expert. But, such warnings aside, the overall impact of the text suggests it is to be taken as more than idle speculation by an armchair dilettante; I believe Geertz wants to be read as Max Weber more than Sir James Frazer. If his argument is worth defending, why does Geertz make no scholarly attempt to do so with the data he found in those villages?

The contrastive approach used by Geertz is rigidified even further in the choice of biographies. "Seeing history in terms of personalities, especially dramatic personalities, is always dangerous," warns Geertz, since paradoxically, or seemingly so, idealized personalities "sum up much more than they ever were."[45] Not heeding his expressed reservations, Geertz chooses a Javanese prince and a Moroccan saint in order to transform men into metaphors. Idealized to a default, these icons stand in for the entire cultural "ethos" of the two "civilizations." For Indonesia the everyman chosen is Sunan Kalidjaga, who is traditionally associated with introducing Islam—without force and reformed for local conditions—into the region. The figure Geertz chooses "to strike off" against the Indonesian prince is Sidi Lahsen Lyusi, an obscure Berber shepherd who founded a maraboutic order. "Two men, two cultures . . ." comments Geertz, adding that "Their differences are apparent, as differences usually are."[46] But the only thing apparent in his textual representation is what Geertz has chosen. Conveniently both "real" men, the kind ethnographers might deal with, are so enshrouded in legend that they can easily be shaped by Geertz to be "more than they ever were." As metaphors of the societies Geertz wants them to contrast, the Indonesian townsman and aristocrat vs. the ordinary rustic Moroccan tribesman, the reader is binarily deluded into thinking this is a meaningful way to essentialize Islam across specific cultures. One is a Yogi and spiritual chameleon, the other a puritanical zealot.[47] Thus, in a simplistic nutshell minus the obligatory comments on similarities, he establishes

how very different Islam turned out to be at the virtual ends of
Islamic expansion. But is this the conclusion or the starting point
for Geertz's analysis? Did these two iconic figures emerge from a
thorough re-search of canonical hagiographies or did they resonate
with meaning from the start simply because they were there and fit
the case for compelling contrasts?

Biographical idealization parallels Geertz's reduction of complex
social and political conditions to a series of "isms" pursued by "ists."
Instead of issues of class, ethnicity, global economic forces, and political ideologies, Geertz explains the civilizational sense of Islam as a
hermeneutical play between labels: scripturalists (who are a little like
fundamentalists) vs. illuminationists and maraboutists (who are not
really like modernists).[48] Geertz further reduces Morrocan Islam to
Fabianism and Indonesian Islam to Utopianism, as though reference
to a second philosophical binary justifies idealizing. Such broad
collective connotations are, as Roy Ellen cogently suggests, "no more
than labels."[49] Fortunately, the term "scripturalist" has not stuck in
recent anthropological discussion of what used to be called Islamic
"fundamentalism." I suspect this is due mainly to Geertz's spin on
scripturalism as either rabidly antithetical to secularism (the supposed
Moroccan case) or a naive remaking of the Quran as a rationalized
deism (the more pragmatic Indonesian approach).[50] By posing the
issue as an intellectual disconnect in which religious dogmatism battles
science and modernity, he loses sight of the political and cultural realities of neocolonialism and global consumerism. Indeed, by privileging
the "guides-for-action side of religious symbols," Geertz's observations on Islam run out of space before getting to the "ordinary behavior" of Muslims relaxing on an international airplane flight.

It all fits neatly, far too neatly. The fit is contrived, quite consciously so, but not in any way approaching a "scientific" sense; even
the loosest standards of social science call for demonstrable proof
rather than interpretive ingenuity. Geertz creates an imaginative
rendering of forms Islam could take, but he offers no "systematic"
evidence that this is in fact what has happened. The problem, as
Munson reminds us, is that Geertz "attempts to reconstruct the
collective imagination of the past without using the texts in which it
is inscribed."[51] Nor does he ask colleagues in history and religious
studies who might be in a position to do so, to so do. Rather, we are
asked to believe that the contrasts elaborated here say something
mega-meaningful about Islam in two "civilizations" as well as about
Islam as such. But what exactly are these civilizations that are so summarily contrasted? Morocco is a "modern" monarchical nation-state
with a king who claims direct descent from the prophet Muhammad.
But Morocco, as Geertz found it, filtered Islam through a sieve most

recently sifted by European imperialist and economic interests. Indonesia is a contested conglomeration of islands merged into a most unwieldy state; there are Muslims throughout the vast stretch of the country, but Geertz had experience with them primarily in Java. Islam is the dominant religion, but it is by no means the only religion; nor do the secular trade politics of twentieth-century Indonesian history mirror the Euro-monitored stewardship of Morocco's royal line.

To state the geographical problem more bluntly, Geertz's knowledge of both areas as an anthropologist is limited to a few villages and towns he visited and lived in.[52] How does such experience, valuable as it is, allow him to speak not only for each country as a whole but for the even grander, and egregiously grandiose, concept of a "civilization"? As Adam Kuper complains, Geertz usually fails to build bridges between the local communities he has studied and the broader regional and national communities.[53] This methodological canard is clearly a primary concern in the lectures, specifically to show that experience in parochial settings could be invaluable for doing macro-sociology or mega-theory of religion. But the rationale for the comparison is backwards; Geertz has chosen Morocco and Indonesia (read Java) simply because he was there in both places, not for any intrinsic comparative purpose borne out of rigorous historical study. Of course, one could compare any two Islamic countries, any two regions within a country, any two villages in a valley, or two alleys in a village. But Geertz never demonstrates that having been there is really the most compelling reason to do so.

Defining Religion Meaningfully

Without further ado, then, a religion is: (1) a system of symbols which acts to (2) establish powerful, persuasive, and long-lasting moods and motivations in men by (3) formulating conceptions of a general order of existence and (4) clothing these conceptions with such an aura of factuality that (5) the moods and motivations seem uniquely realistic.[54]

Clifford Geertz

Geertz was well positioned to present his observations on Islam not only because of ethnographic experience, but as a scholar involved in defining religion as a subject of anthropological inquiry. The canonical *Encyclopedia of the Social Sciences* pegged Geertz as the expert to define "religion." His "Religion as a Cultural System," first written in 1963, has been required reading in anthropology courses on religion ever since.[55] I have always assigned it in my course on the cross-cultural study of religion, despite the nagging realization that most students have little idea what Geertz is talking about. This is especially true of his fivefold definition of religion, for which his article provides an extended, almost rabbinical, commentary. This definition is, by Geertz's admission, a reduction of a premised paradigm: "that sacred symbols function to synthesize a people's ethos—the tone, character, and quality of their life, its moral and aesthetic style and mood—and their worldview—the picture they have of the way things in sheer actuality are, their most comprehensive ideas of order."[56]

For Geertz, the reason for presenting a universalist definition of religion is twofold: anthropology of religion had made no "theoretical advance of major importance" for the previous two decades and it was too self-contained in a "narrowly defined intellectual tradition" of Durkheim, Weber, Freud, and Malinowski. Geertz proposes a definition that would widen or expand on the earlier ideas and reckon with currents of then-contemporary and nonfunction[alism]ing thought. Disarmingly, he warns that although "definitions establish nothing," they can "if they are carefully enough constructed, provide a useful orientation."[57] We are then told that because definitions are explicit, they avoid the problem of substituting rhetoric for argument. That his definition of religion has been so widely cited, even for the purpose of critique, would suggest that definitions invariably do establish something when they circulate widely enough.

To probe this orientation it is useful to approach it, as Geertz himself does, as a "text."[58] What information is given to the reader before the definition of religion is announced with great fanfare? Which premises are the reader asked to accept *sui generis* in Geertz's carefully crafted preamble to his definition? Geertz is quite explicit in stating that there is nothing new out there and what is already out there is narrow and stagnating. Is it really the case that anthropologists had not been writing about religion during and after World War II? The sources cited and ignored by Geertz indicate that some notable anthropologists were concerned with religion.[59] But Geertz argues that none of this led to a theoretical advance of major importance, which may also be read as his opinion that none of these anthropologists were saying what he wanted to say. Such a rhetorical jump is not unique to Geertz; writing three decades after this claim was made, Morton Klass argues that "much of Geertz's criticism is still valid."[60] Why? Because the "theoretical sophistication" in the anthropology of religion has not matched that in the study of social organization and political systems. So does the fault lie with the interpreters of religion or with the complexity of the subject? Is it the case that anthropologists should be able to do for religion what they had apparently done so well for social organization? And, who had decided that the "transcendent figures" of Durkheim, Weber, Freud, or Malinowski narrowed the chance for advance in the anthropological study of religion? Was it really blind faith in past idols of the field that explains, as Geertz implies, the seeming lack of anything new and sophisticated over a twenty-year period in anthropological study of religion? There was, of course, a rather significant world war in the middle of this gap; surely this might have given pause to reconsider what it meant to study religion.

If there was indeed a crisis, a need to revivify the "dead hand of competence,"[61] caution was required. To move beyond what Geertz himself acknowledges as valuable contributions by competent scholars, it is important to avoid "arbitrary eclecticism, superficial theory-mongering, and sheer intellectual confusion."[62] The rest of the article, this must imply, would attempt to overcome these substantive obstacles. Geertz narrows his own focus to "the cultural dimension of religious analysis," a phrase borrowed from Parsons and Shils, two of the intellectual influences lauded in the introduction to *Islam Observed*.[63] He is quick to note that "culture" has been unjustly maligned by British social anthropologists, although he—parenthetically in this case—does not see the term as having any "unusual ambiguity." This sets the stage for the Geertzian spin on just what "culture" is.[64]

Since religion is a "cultural" system, understanding how Geertz interprets this preeminent concept in American anthropology is necessary before moving on to religion. Culture, Geertz explains here, "denotes an historically transmitted pattern of meanings embodied in symbols, a system of inherited conceptions expressed in symbolic forms by means of which men communicate, perpetuate, and develop their knowledge about and attitudes toward life."[65] Having discussed this in other contexts, the definition is given here without rhetorical need to defend it or convince the reader that it fits into the rather vast literature on culture, especially in the previous two decades when anthropologists were apparently unable to make similar advances in defining religion. To explicate this definition, it is useful to follow Geertz's own well-versed exegetical method, without further ado, and divide it into analyzable segments. So *culture (1) denotes an historically transmitted pattern of meanings embodied in symbols, (2) a system of inherited conceptions expressed in symbolic forms (3) by means of which men communicate, perpetuate, and develop (4) their knowledge about and attitudes toward life*.

denotes an historically transmitted pattern of meanings embodied in symbols

Geertz chooses to "denote" rather than "connote" culture, a curious choice for someone who thinks that definitions establish nothing.

Words that denote are meant—by definition, of course—to establish
everything. "In denotation," declares a basic college text on literary
analysis from the time Geertz was writing, "one can believe what
he reads; there is no ambiguity, or at least there should be none, in
the particular meanings of the words."[66] But what could be more
ambiguous than a "pattern of meanings" and how can such a notion
be "embodied in symbols" without conjuring up the theological
image of the word made flesh? These are metaphors, perhaps necessary for as ambiguous a category as "culture," but they connote with
poetic license and not with scientific precision. This "pattern" of
meanings for Geertz is certainly not a thing of "shreds and patches,"
but the choice of wording here reflects an American fetish with
patternizing, and thus patronizing, culture, as exemplified in anthropologist Ruth Benedict's eloquent but unusually ambiguous *Patterns
of Culture*.[67]

    A pattern of "meanings" assumes a logic or grammar to human
social behavior, at the very least a texture to the textualization. While
Geertz himself chooses a literary or "textual" approach as opposed to
a linguistic model, culture is still reduced to a nonmaterial shared
consciousness, echoes of the superorganic debate engaged by such
passing notables of the day as Alfred Kroeber. By "historically transmitted" Geertz assumes that culture does not die with individuals but
is carried on from one generation to another through what an earlier
generation would have unambiguously called customs and traits.
Unstated, but still present, is the unchallenged assumption that the
anthropologist is the one to decode this meaning. Geertz is, of
course, a vocal proponent of seeing things from the native's point of
view, but it is his own reading and his contextualizing of that point
of view that invariably re-presents their views.

    The Geertzian twist, his major contribution to the study of culture,
is in the argument that these patterns are "embodied in symbols."
This critical phrase is also the most ambiguous and un-denotative part
of the whole definition. The relevant dictionary rendering of "embody"
is to give "a material or discernible form to (an abstract principle,
concept, etc.)."[68] In order to imagine culture in a way discernible to
our commonly shared senses, it is necessary to look not simply at
what people do and how they do it but at the shared symbols that
connote—necessarily so if "symbol" is to mean anything—the meaning of what people do and say. Yet, it is the ethnographer who
materializes the embodiment of fuzzy conceptual entities such as
symbols. Geertz does not want to reduce the cockfight, for example,
to how this activity functions in a social sense nor how it serves political ends but rather what it really "means." The cock as symbol is thus

the stuff of culture; leave flesh-and-engorged-blood cocks to veterinarians and gigolos. But, as Crapanzano cogently reminds us, "Cockfights are surely cockfights for the Balinese—and not images, fictions, models, and metaphors."[69] Symbols are only capable of embodying something metaphorically. The "body" embodied is akin to a sculpture, since it has no "body" of its own apart from the meanings created for it, or the image of a shape shifter; the real beauty of its meaning is so much in the eyes of the beholder. As Talal Asad points out, Geertz uses "symbol" at times for an aspect of reality and at times for its representation.[70] If Geertz sees no "unusual ambiguity" in such a notion, he is very much a victim of his own symbolizing.[71]

a system of inherited conceptions expressed in symbolic forms

The second phrase shifts from aesthetic terms—patterns, meanings, and symbols—to a more scientifically inclined rendering of the same ambiguity. A "system" is decidedly more analyzable than a "pattern." Indeed, a systematic approach only makes sense if there is a discernible structure to be accessed. Culture as a system is not something that is historically transmitted but rather "inherited," switching to a biological metaphor. Geertz replaces his "meanings" with "conceptions," a more fertile scientific phrasing. Reverting to the standard dictionary rendering once again, "conception" refers to "that which is conceived" but this may be something material (as in an embryo) or a mere mental notion, even one of fancy. Indeed, the English verbal form, to conceive, can be used for either the physical or mental sense of creating. It is precisely this dual looseness that allows Geertz to clothe his definition of culture in more acceptably scientific language. These inheritable conceptions are thus "expressed" in symbols. Once again it is not the material or behavioral aspects of culture that have meaning, but rather the way they are symbolically perceived, although Geertz alters "symbols" to the more scientifically correct metaphor of "symbolic forms." Apply such a muddled idea to the conception of actual birth: it would not be the biological features that determined whether a baby was male or female, but how that baby was perceived. Gender, in this expressed sense, would overrule sex. This is not to deny the importance of symbolism or meaning, anymore than to ignore the social construction of gender, but is reducing culture to what it "means" to observers in any way "scientific" and by any significant measure "systematic"?[72]

Throughout his writing Geertz cavalierly shifts from literary to scientific metaphors as if it were perfectly natural and sensible to do so.

In defining "culture patterns," he draws a direct analogy to genetics.[73] First, "unlike genes" these culture patterns are extrinsic or outside the organism—culture survives the death of individuals—and then "like genes—they provide a blueprint or template." This sets up the following rhetorical maneuver:

> As the order of bases in a strand of DNA forms a coded program, a set of instructions, or a recipe, for the synthesis of the structurally complex proteins which shape organic functioning, so culture patterns provide such programs for the institution of the social and psychological processes which shape public behavior.[74]

Quite cleverly, Geertz first oversimplifies DNA by amplifying "coded program" to a generic "set of instructions" or eminently cultural "recipe." Biologists use such terms to communicate a rather complex biological process in terms ordinary people might follow, but metaphorically duplicating DNA is hardly a scientific explanation. Having put the scientific concept into accessible laymen's terms, Geertz adds that this is how cultural patterns work as well. But the analogy is stretched beyond utility. DNA is decodable because scientists have the means to identify chemical bases; there is no such indisputable chemistry in the study of culture. The analogy is a better fit for alchemy, the appropriation, inappropriate in this case, of scientific language to formulate fools' gold.

by means of which men communicate, perpetuate, and develop

A judicious parsing of Geertz's elongated sentence would suggest that the "which" here refers back to the "pattern" or "system" that is either transmitted or inherited in symbols or symbolic forms over time among people. The assumption here is that men, and one assumes women, are able to communicate, keep on communicating, and communicate something new because of such an underlying pattern or system. Yet if the pattern or system is to function as the means for this communicating, there must be a shared grammar. It would be absurd to argue otherwise for language, the distinctive form of most human communication. The grammar need not be articulated by those who speak a language, but all languages are capable of being reduced to logical grammars. Thus, by analogy, there must be a grammar informing "culture" as such, either for individual societies or human existence. While I do not deny the attraction of the metaphor, it is nevertheless circular. We know by experience how languages function; our linguistic models are designed to reduce specific language acts to basic principles that seemingly underlie the human capacity for speech. For the pragmatic purpose of systematic linguistic study, languages exist; cultures, despite the compelling metaphorical link to language, are merely posited. To say we can analyze culture the way we do language is tantamount to saying we could study angels the way we do human physiology. Whether angels exist or not, the human metaphor works because that is how they have been symbolically personified. Is there any logical improvement with Geertz's approach to culture?

Speaking of angels, it is relevant to note the created form a little beneath them, namely "men." Geertz constructed this definition of culture at a time when academic prose consciously and unconsciously substituted "men" for "humans." There was, at the time, little awareness that such a mundane linguistic habit perpetuated a profoundly male bias in anthropological writing. The Geertzian corpus is very much a male domain. Women exist, but often drop out of the picture as "real" individuals as soon as Geertz finds an interesting symbol to textualize. Thus, Geertz and his wife Hildred both arrived as anthropologists to a Balinese village in 1958, but she disappears from sight as soon as the introductory trope moves beyond the "gust-of-the-wind stage" of his field entry.[75] Women as culture bearers appear from time to time, but usually because of what they symbolize or what they wear.[76] Some critics, such as historian Kathleen Biddick, read Geertz's proclivity for meaningfully penetrating social events like Balinese cockfights as evidence for a "rhetorical collusion" with the conventions of a hyper-masculine genre.[77] Ah, men. Amen.

their knowledge about and attitudes toward life

This sequential unraveling of what culture might mean leads us ultimately, in terms of the definition provided, to "life" as such. There is a pattern or system, embodied or expressed through symbols, which allows for the communication of "knowledge" and "attitudes." The drift of Geertz's "textual attitude," as Edward Said might call it, is that cultural knowledge is "about" life. As the article goes on to explore, Geertz argues that culture is really about important matters, like the meaning of evil, rather than, for example, Trobriand canoe building. It is a plea to concentrate on those items of knowledge that are meaningful, at least to the anthropologist, about culture. Equally important here are attitudes "toward" life, a vague reference to what might also be called a worldview or philosophy. Knowledge, for the anthropologist, has always been quantifiable and collectible. There is no dearth of ethnographic data, especially in the expansive Human

Relations Area Files at Yale University, and little reluctance by anthropologists to use such data for theorizing about various aspects of society or culture. Attitudes, on the other hand, are notoriously difficult to identify in anything approaching a scientific spirit. It is relatively easy to describe in a thick ethnographic tome how a canoe is built; it is a different matter to explain why it is built beyond the fact that it serves a useful purpose or what canoes seem to symbolize in the language of the people who use them. Anthropologist routinely do both, but only the former tends to go unchallenged and, admittedly, often unread.

One way of contextualizing Geertz's definition of culture is to look at what is not mentioned. Geertz describes culture minus the social; missing here is any specific reference to social organization, economic systems, or political institutions. One could pursue culture, as far as this definition goes, without resort to kinship obligations or factional fissioning. Nor is there any specific need to conceptualize culture from the nitty gritty of ethnographic fieldwork. I could imagine a historian, philosopher, visionary scientist, or even theologian devising a similarly broad and generic definition. This is not to suggest that Geertz was unaware of what anthropologists uniquely do in their study of human society, nor that he under appreciated this role. However, in his desire to offer a definition palatable outside his discipline, he in effect dismisses the central methodological tools anthropology can offer to the debate. There is room to build on the shoulders of men like Franz Boas and Bronislaw Malinowski, as well as women like Ruth Benedict and Margaret Mead, without ignoring their substantive contributions as ethnographers engaged in the thick of cultural interaction.

Having generalized "culture" in a way that could theoretically appeal to almost anyone, Geertz moves from this paradigm about culture to his definition of religion. Although the reader may not be aware of the ghosts that haunt his entry into this subject, Geertz is clearly at pains to distance himself from established starting points in the way anthropologists previously defined religion. There is no echo here of Tylor's minimum definition of religion as "belief in Spiritual Beings."[78] For Geertz, humans are enmeshed in a Weberian web of symbols of their own making. A scientific approach to religion presumably holds such notions as spirits and gods in limbo as nonmaterial objects unsuitable for empirical investigation. But Geertz does not wish to align himself with the agnostic predetermination of Durkheim, who limits ultimate meaning to the social function of ritual and the social relevance of a world so eminently divisible into sacred and profane. Religion, for Geertz, is just another human way of looking at "ethos" and "worldview." Both these terms, widely shared outside anthropology at the time and today, intellectualize religion as a philosophy of life. Whatever each is said to mean, the bottom line is that the claim for a transcendent supernatural must be held in abeyance for scientific analysis to take place. Religion becomes a projection, whether in a strict Freudian sense or not, and politically manipulative, even if one is not a card-playing Machiavellian nor a card-carrying Marxist. For anthropologists like Geertz, however, religion is not approachable as a viable statement of truth.

While eschewing a theological approach to religion, Geertz in fact appropriates the theory (hermeneutics) and method (exegetical commentary of a defined sacred text) common among theologians at the time. Talal Asad thinks Geertz inadvertently takes up the standpoint of theology: "This happens when he insists on the primacy of meaning without regard to the processes by which meanings are constructed."[79] The strong appeal of Geertz in the humanities is precisely because "he speaks fluent hermeneutics."[80] Geertz as secular theologian reframes classic sociological explanations of religion into the dialect of those scholars he appears to want most to impress. Consider how he moves from Malinowski's interpretation of religion as a way to relieve emotional stress via an ethnographic case study from Africa, plucked from Radin's *Primitive Man as Philosopher* rather than his own ethnographic fieldwork to the "problem of suffering" and ultimately to the "problem of evil." If a definition of religion has no place for spirits or God, how can it comfortably address the problem of "evil," a theologically driven teleological trope if ever there was one?

We are now ready to look at Geertz's quotable five-pillars definition of religion. Recognizing the imprecision prevalent in the usage of "symbol," Geertz begins by defending his sense of it as a "vehicle for conception," the "conception" being the meaning of the symbol.[81] With this explication there is little that could not conceivably be a symbol. Not only are humans enmeshed in a web of meaning, there is no escape; the web might as well be metaphorized as an expansive ether. The choice of vehicle here and the insistence that symbols are somehow tangible lead Geertz to commit the metaphysical blunder of confusing what reality should mean with what reality must be. To be sure, Geertz is not an advocate for the tangibility of "social facts"; yet, he is confident that symbolic forms are not merely mental or idle speculation but rather as "observable" through public social acts as a marriage ceremony or agriculture. The rhetoric here leads us astray from the start. Surely we can only meaningfully use the term "symbol" for what we conceive, not observe, to be the case. The numeral 6 and the Christian cross, Geertz suggests, are both symbolic "because they are tangible formations of notions, abstractions from experience fixed in perceptible forms, concrete embodiments of ideas, attitudes, judgments, longings, or beliefs."[82] Yes, they no doubt are, but these are symbols that Geertz cites precisely because they are recognizable as having a meaningful symbolic history to both the author and the reader. However, who makes the symbolism of a cockfight tangible, who abstracts it, who embodies it for the reader? Geertz would say he is merely reading the textual play of life over the shoulders of his informants, but how are we to know his reading is the same as their living? It is hardly novel to say that social events have symbolic meaning; the problem is, as always, in the representation.

To return to the parsing of this definition, religion, like culture, is identified as a "system," indeed, in the title of his article, as a "cultural system." Geertz goes on to expound on this sense with a number of synonyms that convey quite different connotations. The system of symbols is also a "complex" of symbols or a "cultural pattern." To further clarify—or perhaps confuse—what this means, we are introduced to Geertz's understanding of "model," meaning in this case a "set of symbols." But there is quite a denotative leap of faith from system to model. The Solar System, at least since Copernicus, is thought by most people to really exist; a Ptolemaic, Copernican, or Einsteinian model of the Solar System is hopefully not to be confused with what it is meant to explain. So, does Geertz want us to think of religion as an existing system in real time or a model of what that system, or whatever it is, might be were the supernatural spirits religious people say exist, really to exist?

Rather than address this epistemological issue, Geertz teases us with an intriguing classification of models "of" versus models "for."[83] A model "of" here is like a flowchart for building a dam or, one might add, the diagram of the double helix for modeling DNA. It allows us to visualize something we take to be real by making it "apprehensible" to our senses. The point is that such a model is a way of symbolically rendering what is essentially nonsymbolic. A model "for" reality is said to do something different. Theoretically, this means "the manipulation of the nonsymbolic systems in terms of the relationships expressed in the symbolic." So when you actually construct a dam according to the conclusions to be drawn from a flow chart, you are operationalizing a model "for." In this case the model "of" would seem to always precede the model "for" in the sense that builders need blueprints, even if only in their minds. But we are also told that genes, by analogy, are models "for" reality as though our scientific modeling on genetic structure can substitute for the actual biochemical processes of reproduction. This muddled modeling by

Geertz is further mystified by the assertion that in culture and religion modeling "of" and "for" somehow merge so that one can become a "mere transposition" of the other.[84] It could, I suppose, be argued that his entire article is both a model "of" religion—at least for Geertz—and a model "for" for those who see it as paradigmatic, but although it is often cited as a blueprint there are few structural studies of religion actually built on the model of his definition.[85]

This idiosyncratic use of "models" deflects and at the same time highlights an essential problem with Geertz's articulated approach.[86] A model is not a system, nor is it synonymous with a pattern or complex. Models may be systematic, patterned and complex, but they are designed to represent or, quite literally, "model" an assumed reality, not metaphorically take its place. We assume a "reality" to genetics that our modeling of genes and DNA is meant to understand. The DNA model is an explanatory model "of" that reality, but certainly not a recipe "for" making genes. The hydraulic flow chart of how a dam works is a model "of" a dam, to be sure, but it is of damn little value if it is not actually used to build a dam. Or, to use a more mundane example, the instructions for building a model rocket can both show how the parts relate to each other and how to put them together. Whether you see this as a model "of" or "for" depends on your purpose. The words used for the system, pattern, complex, and model of symbols that Geertz sets out as the base of religion are the only things tangible. But what is the comparable reality that is modeled in religion? If you are prepared to accept spirits and angels as equivalent to chemical strings of DNA, Geertz's model poses no dilemma. Scientists, even those most skeptical of reductionist materialism, would have little to gain from such a model.

That Geertz's equation of system with model is flawed is further shown by his assertion that this "system of symbols" "acts." Linguistically, it would seem this should mean that it produces an effect or influences something else. A scientist could note that when the temperature falls below freezing, it acts to freeze water. A functional sociologist might say that performance of a group ritual acts to promote solidarity. A theologian might say that the Devil acts to influence people to do evil. It is not clear, from Geertz's figurative usage here, in which of these ways his booking of "acts" is to be read. Between freezing and the Devil is one hell of an epistemological gap. The result of this acting is the establishment of "moods" and "motivations" in men, so perhaps the best way to see through the rhetoric is to start from there. It appears that Geertz is applying the extended dictionary sense of "mood" as the "atmosphere or pervading tone of a place, event, composition, etc.," which would circle it back to his paradigmatic premise of "ethos" as the basis of culture. But how does one establish criteria for defining such a "mood" in a sense beyond the psychology of individuals? A man or woman can have a describable mood, but does a system, village, a society, a culture, a civilization? Or, how does one escape the equally viable English usage of "mood" to describe a temporary state of mind or feelings? And for the "motivation," is it conscious or unconscious? Does Geertz really expect us to believe that moods and motivations are as observable as marriage rites and agricultural practices?

The key activating factor in the definition becomes "formulating conceptions of a general order of existence." The moods and motivations, whatever they are, are established through "conceptions." This apparent [double]cross between a logos doctrine and a philosophical black box leads us back to the same problem noted with Geertz's definition of culture. How is the anthropologist, or any student of religion, supposed to recognize and document a "conception"? Is this based on what a specific man or woman says, what many people say, what statistically most people say, what people say other people say, what is written down, or what an anthropologist not entirely fluent in the local language interprets as being the case? The conception here, in a religious sense, is a model of, and apparently not for, "a general order of existence." This is, I suspect, a tremendous let-down for most theologians reading through Geertz's narrative. Forget about divinity, ritual, passion plays, and institutions; religion is just the general way people make sense of the world around them. The *Heilige Geist* is effectively exorcised from the *Geistgeschichte*.[87] How this differs, except in minor metaphorical ways, from the bland but widely traveled "worldview" is hard to understand. If Geertz's point is to argue that religion, like art, ideology or common sense, is just another cultural system, why bother?[88] After all, it would just be common sense.

His extended exegesis of the definition suggests that this is not his intent, that he wishes to tread the same ground as theologians and philosophers with his own anthropological model "for." This is clearly the point of his summary statement:

> The anthropological study of religion is therefore a two-stage operation: first, an analysis of the system of meanings embodied in the symbols which make up the religion proper, and, second, the relating of these systems to social-structural and psychological processes.[89]

The problem, for the reader, is that only the first of these operations is laid out in the article. Having defined religion and clothed it with a variety of ethnographic examples, Geertz only haphazardly fulfills the ultimate—or is it secondary—goal of relating symbols back to the "socio-structural and psychological processes" from which they were derived in the first place. The system that Geertz spins is left in suspension.

Take the Bororo *arara*, the portentous Amazon bird that scholars of an alleged primitive mindset so long assumed to fly according to a contradictory law of its own antinature.[90] Geertz cites this "parakeet" as an example of how religion makes sense out of social needs:

> A man who says he is a parakeet is, if he says it in normal conversation, saying that, as myth and ritual demonstrate, he is shot through with parakeetness and that this religious fact has some crucial social implications—we parakeets must stick together, not marry one another, not eat mundane parakeets, and so on, for to do otherwise is to act against the grain of the whole universe.[91]

The Bororo man, suggests Geertz, is not foolish enough to believe he is the same as a "real" parakeet; hence a Bororo man does not climb trees to mate with one. So the issue is clearly symbolic, the meaning of parakeetness for the Bororo paralleling the meaning of a cockfight for the Balinese. Such rhetoric, we are told, has a power to convince the speaker and those around him in a "supremely practical way," given the way things "really are." This is not Geertz speaking as an ethnographer among the Bororo or even reading an ethnographic text about the Bororo; he is adding his own spin to a long-standing and academically supercilious debate over how primitives think. The *arara* clearly means something in Bororo religion, but when a particular Bororo is on record, indirectly in this case, as saying he "is" a Bororo, the problem should begin with what the meaning of "is" is.[92]

Critics of Geertz fault his interpretive mode for distancing from earlier functional explanations and privileging symbols as something good to think rather than byproducts of the political and economic acts "real" people engage in all the time. The model provided of religion is a hermeneutic in need of grounding in the nitty gritty of daily life. "How does power create religion?" asks Talal Asad.[93] It is one thing to say that religious symbols are socially powerful, but this is a point that needs to be investigated rather than reinvented. The historical use and abuse to which the Christian cross has been put by the KKK says more about the relationship between political agendas and religious justification than any statement a Bororo man might have made to a German traveler more than a century ago. What is lacking in Geertz's well-meaning definitions of culture and religion is how to turn the symbols, as represented by the anthropologist, back on the society where they have practical currency. Many of us who read *Islam Observed* find the same lack there.

Where did the Observed Muslims Go?

Geertz's study, focusing on the social conditions of the two communities at the time of conversion and at the present time, is able to delineate the variety of responses possible to the same socio-religious force, viz., Islam. But, if one were to study a window and a brick wall after a rock was thrown at each of them, and only looked at the window and the wall, one might learn much about the nature of brick and glass, but one could hardly call the results of the analysis "rocks observed."[94]

David D. Laitin

Clifford Geertz was invited to give the Terry Lectures not because he was an expert in Islamic history, a philosopher in the old disciplined sense of religion, or a Muslim himself. He came as an established anthropological scholar who had the unique and valuable experience of having lived among and studied Muslims in two differing cultural contexts. Ironically, Geertz mounted the podium because of where he had been, but proceeded to outline a view of Islam largely based on where others before him had been or at least where he wished they had been. The starting point is that fieldwork was formative in his interpretation of Islam. "The bulk of what I have eventually seen (or thought I have seen) in the broad sweep of social history I have seen (or thought I have seen) first in the narrow confines of country towns and peasant villages."[95] As I see it (or think I see it), this can be interpreted in two ways. The fieldwork experience in an almost alchemical sense could shape the way in which an anthropologist views the other. Ideally, this should lead to sifting through an analyst's inevitable ethnocentric assumptions and appreciating beyond the obvious differences that Edward Tylor so long ago dubbed the "psychic unity of mankind." Another way of seeing this is that specific data or information learned or collected in fieldwork provide the kind of documentable evidence needed to flesh out or flunk out specific models and theories of or for culture and religion. The first nuance speaks to an anthropologist's authority, while the second allows someone else to determine his or her credibility.

We know, because Geertz is anxious that we do know, that he did his share of ethnographic site-seeing in real villages and towns rather than conceiving a thick description from casual conversations or formal texts. His *Religion in Java* and analysis of the Moroccan bazaar include recognizable information from informants. But the odd thing about *Islam Observed* is that we never actually hear the words, even through the anthropologist's representation, of the Muslims he obviously spoke with. Since Geertz has produced competent and thick ethnographic description in other texts, I am confused by the lack of direct references to Muslim informants in a study that purports to "observe" Islam. The notion of "observing" Islam is a telling oxymoron, especially as the title of the seminal text for the anthropology of Islam. In the field only Muslims can be observed; "Islam," whether in the abstract sense of a religion or the common-sense notion proposed by Geertz, can only be represented.[96] Since Muslims might as well be unobserved for the purposes of *Islam Observed*, Geertz's leap of hermeneutic faith results in an Islam obscured.

In the village alleyways Geertz may have talked to men with the likely names of Muhammad and 'Ali, but in *Islam Observed* the individual Muslims encountered are mostly icons. Ironically, Geertz at times reconstructs imagined dialogue of historic personages as though he is transcribing from a taped conversation.[97] His "two men, two cultures" approach reduces two very complex cultural contexts to two dead and legendized men he never met. Whether or not Geertz adequately represents the lives of these two men, why is it that the many live men he had the opportunity to talk with do not appear relevant for interpreting the cultures they actually lived in at the time? Indonesians and Moroccans appear in the aggregate from the beginning, but only at the very end, as an aside, do real Muslims suddenly seem relevant.[98] Geertz leaves the reader of *Islam Observed* with a frightened Moroccan student, flying to an American university "with the Koran gripped in one hand and a glass of Scotch in the other."[99] Contrasted with this rather backsliden "maraboutist" is an Indonesian student condescendingly described as "one of the country's few promising scientists," who—if he does not get a chance to "build their bomb"—will spend his life working out an "almost cabalistic scheme" merging the truths of physics and religion. If these are the most relevant Muslims that Geertz observed in all those years of fieldwork, the case for the relevance of ethnography has not been made.

While much discussion of *Islam Observed* has focused on Geertz's distinctive interpretive approach and the accuracy of his knowledge of Islam, surprisingly little attention has been drawn to the missing informants.[100] The thinness of ethnographic data in his comparative study, a function of his own choice, is seldom raised as a pertinent issue. Daniel Pals, for example, notes that this text "does not offer a crisp logical argument in defense of a definite thesis about religion," but then glosses over this fault by referring to Geertz's "keen interest in the particularity of each culture he interprets."[101] Non-anthropologists, as in the case of Pals, tend to interpret Geertz's textual celebration of "particularity" as ethnography of "high quality," as though the valorization of ethnographic data, even when not described, is enough.

Islam Observed became a pivotal text not because it applies ethnographic observation to Islam but rather because a recognized authority in anthropology was consciously and eloquently reaching across disciplines at a time when such academic boundaries were already in flux. Do not dismiss us as dealers in exotica, Geertz asks his audience; we anthropologists are not so dense as to assume the Islam of one village stands for the whole tradition. But, like historians, political scientists, sociologists or economists, we are trying to sort out the link between parochial understandings and comprehensive ones.[102] Proving this point, or attempting to prove it, Geertz analyzes Islam using information primarily derived from historical sources rather than the specific conversations from "wandering about rice terraces or blacksmith shops talking to this farmer and that artisan."[103] If the insights of these village philosophers are not worth repeating, ethnography comes across as little more than a gambit. Rather than a view from the bottom up, Geertz positions himself to give a "distant and patronizing perspective."[104]

Returning almost empty-handed to those rice terraces and blacksmith shops, Geertz frames his observations along a historical timeline. After noting Geertz's emphasis on the formative period of Islam's introduction to Morocco, Dale Eickelman observes that "the lack of sources that reveal the sociological nature of this period limits its possibilities for a comparative study."[105] I find it ironic that Geertz spends so much time on the formal history of Morocco and Indonesia when he rejects the British functionalist dictum repeated by Evans-Pritchard that anthropology is history or it is nothing. Here, in a forum where he was invited to show his wares as an ethnographer, it is his interpretation of an unobservable history, essentialized as it is, that drives the argument.

Back to the Field[work]

"Islam," without referring it to the facets of a system of which it is a part, does not exist. Put another way, the utility of the concept "Islam" as a predefined religion with its supreme "truth" is extremely limited in anthropological analysis.[106]

Abdel Hamid El Zein

The irony with *Islam Observed* is that it legitimizes anthropological study of Islam to many non-anthropologists but serves poorly as either a model of or for the anthropological analysis of ethnographic data to better understand the socially expressed dimensions of Islam. Historians of religion credit Geertz with drawing attention to the importance of culture and culturally significant symbols.[107] Indeed, quoting Geertz has been so common-place, it is the odd book on religion that does not refer to him. This is not surprising, since his open flirtation with hermeneutics and phenomenology fits well in the humanities perspective informing much of religious and theological studies. By studiously avoiding Durkheim and the whole gamut of British functionalism, such an approach is understandably palatable to disciplines with no tradition and little interest in observing individual Muslims. But in pursuing the idea that religion can be reduced to conceptions of a general order of existence, Geertz elides the major sociological truism that it is also about conceptions of society and the behavior based on these conceptions. "Different conceptions of Islam are consequently not simply different explanations of the universe per se," explains Robert Launay, "but rather different ideologies of the social universe."[108]

Geertz cut his methodological teeth on local "islams," in Abdel Hamid el-Zein's well-intentioned sense, and then set his sights on "Islam" as such, the assumed target of all the local variants. Meanwhile, while a generation of students has memorized how Geertz defined religion, a host of ethnographers has been observing Muslims wherever they might be found. As a result, thanks in part to Geertz himself, the anthropology of Islam is no longer a default "Oriental" or "Middle Eastern" specialty. While a specific "anthropology of Islam" seems always to be fixed in prolegomena still-birth, ethnography in Islamic settings has contributed a substantial literature yet to be surveyed for its girth. There is no dearth of published analysis on how Islam is experienced in particular social settings, qualitative judgments notwithstanding. Some ethnographers have added reading skills of formal Islamic texts, thus combining on-the-ground participant observation with literary study of the very texts that symbolize Islamic tradition; others have tackled controversial political aspects in contemporary Muslim groups.

Reflexivist critique of ethnographic method and inevitable debate over theoretical stance and circumstance have hardly hindered the ethnographic study of Muslims. Such criticism, provided it eventually escapes the solipsist rut through which academic discipline tends to channel it, serves as a valuable corrective to dogmatic stagnation. But theory that does not respond to the exigencies of actually applying

methods soon fades into dueling rhetoric over the latest, or most
recently uncorked, intellectual fads. Those few anthropologists, like
Geertz, who have moved successfully to center stage with philosophers of religion and literary critics, are to be commended. But it is
still true that the ethnographer who enters a parlor populated largely
by other brands of intellectuals must generally leave his walking-
through-the-village shoes at the door. Geertz waltzed in with
Islam Obscured, but he knew better than to bring any real villagers
with him.

If he had, I wonder what they would have thought about the
following Geertzian aphoristic truth:

> Whatever else "Islam"—maraboutic, illuminationist, or scripturalist—
> does for those who are able to adopt it, it surely renders life less outra-
> geous to plain reason and less contrary to common sense. It renders
> the strange familiar, the paradoxical logical, the anomalous, given the
> recognized, if eccentric, ways of Allah, natural.[109]

Since Geertz assumes culture should be reduced to a commonly
sensed meaning, such sentiment is not surprising. But whatever the
"whatever else" might be, this is surely what the lived experience of
most Muslims is *not* about except in an abstract model that is often
impossible to follow. I doubt that this sentiment was ever expressed
directly to Geertz by a Muslim on a rice terrace or in a blacksmith
shop. Not only does it not represent a native point of view, it invents
it in terms that fit the author's preconceived sense of what religion as
such should be. Negotiating the given meaning of symbols, forging
ahead through doubt, questioning what other people think, fretting
over what the "truth" might be: surely these are the mundane facets
of living in a world that religious beliefs, dogmas, officials, and institutions often make *more* rather than less confusing. Indeed, Geertz's
desire to package a comparable "common sense" of religion leads him
to miss the compelling attraction of paradox and mystery missing in
otherwise ordinary life. I suspect that a Muslim would be far more
likely to tell Geertz that the ways of Allah are always natural; humans
are the eccentric ones. For a Muslim suicide bomber, to take an
extreme case, Islam does not make the extraordinary act of killing
oneself "natural"; the whole point is that it makes an otherwise
unnatural act seem supernatural.

What would the audience for the Terry Lectures have said if
Geertz had thrown away his carefully crafted lecturese and reported
instead the actual words of Muslims he talked with in Java
and Morocco? How embarrassing would it have been if Geertz had

admitted that many of the Muslims he talked to had contradictory ideas and did not seem to have a proper hold on Islam with a capital "I"? What would the learned Yale humanities professors have thought of this anthropologist, this marginalized intellectual by disciplinal default, this man with the dirt of everyday life still caked on his well-worn shoes? Would they have walked out in disgust if he had paid no attention to recorded history, if he failed to eulogize and idealize the elite saints and sultans? Would some Orientalist, with seven languages swimming in his head, have shouted out "That's not the Islam in the Quran; that's just some ignorant peasant who can't even recite the *fatiha!*" Would there ever have been a book called *Islam Observed?* How exactly does the comparative study of religion suffer when the people who live the religion on a day-by-day basis are consulted? By leaving the ethnography out, these questions are not even raised, let alone resolved.

I suggest, only half facetiously, that anthropologists take a good look at their anthropological selves in the mirror and face a reality that other disciplines tend to see about us much more clearly. Anthropology in its cultural sense is ethnography, going to the locals for our data, or it is nothing different. Scholars in many disciplines interpret "Islam" and most Muslims of necessity look for the meaning of their faith without any outside prodding. It is not that the anthropologist cannot master another field of study, but I know of no other discipline that mandates, theoretically in some recent cases, participant observation as "the" defining method of data collection. As el-Zein knew so well, it would be foolhardy to look in the field for "Islam" in that essentialized and decontextualized sense so many scholars want to define. As an ethnographer who has seen context, why would you want to? There are plenty of Muslims out there—some good, some bad, some indifferent—depending on who is doing the interpreting. Anyone patient enough to read through ethnographic reports, even the dry-as-dust-variety that Clifford Geertz long ago forgot how to write, will find a wide variety of "islams" and all sides of a lively debate over who is a Muslim, what is *haram* or *halal,* and what Muslims should be doing when they are also conservatives or communists, male or female, young or old, rich or poor, in a good mood or cruelly motivated.

Chapter 2

Ernest Gellner: Idealized to a Fault

And is it not significant that when social anthropologists burrow in the micro-structures of Muslim societies, they generally come back with a picture very compatible with that of Ibn Khaldun?[1]

Ernest Gellner

If there is any one book by an anthropologist purporting to explain Islam or Muslim Society that should be avoided because it is so summarily patched together and indignantly indifferent to available scholarship, that text could easily be Ernest Gellner's *Muslim Society*, which appeared in the Cambridge Studies in Social Anthropology series in 1981. One reason this book does not hold together is that the twelve chapters are mostly previously published articles and book reviews from 1963 to 1979. Despite a lengthy first chapter stating many of the themes to be replicated in not very dissimilar variation throughout the compilation, the primary rationale for the text as a whole appears to be that these themes were "gestating for over a quarter of a century."[2] While I have no academic death wish to impugn the reputation of Gellner as one of the preeminent British social anthropologists of his generation, nor his profound influence as a teacher and mentor, I do find his approach to understanding and writing about Islam severely flawed; *Muslim Society* serves as an ideal [*sic*] lesson for what not to do in anthropological analysis of Islam.[3]

Clifford Geertz, an itinerant critic of Gellnerian reason, classifies his British colleague as one of those anthropologists who tries "to save us from ourselves."[4] My goal is more modest: to salvage the anthropology of Islam from an essentialized philosophical dead-end. I propose in this chapter to proffer a response to Gellner's rhetorical question about whether or not it is significant that social anthropologists seem to see things in a way compatible with the noted, late fourteenth-century, Arab scholar known as Ibn Khaldun. Along the way it will be necessary to examine the hyperbolic assertion by Gellner, among others, that this Ibn Khaldun is "the greatest sociologist of Islam."[5] My own *muqaddima* introduces the context of Ibn Khaldun, a precocious precursor *par excellence*, moving on to the scholarly discourse that has evolved largely outside the circle of those who actually read Arabic or study similar historical texts. Revisiting what Ibn Khaldun was saying so long ago and so far away is critical to assessing the main theoretical thrust of Gellner's interpretation of Muslim society in which Khaldunian "sociology" along with David Hume's "enlightened" views on the origins of religion, Emile "social cohesion" Durkheim, and Max "idealized to the maximum" Weber are processed into a theoretical muddle. Having dealt with the core of the argument first, we can proceed to Gellner's "flux and reflux" theory through an idealized Islam. In a particularistic sense, Gellner's text says little about Islam but much about how brilliant scholars can veer off into sheer folly when they privilege presumptions and venture outside areas in which they have established a credible, even if contentious, expertise.

A theoretically trenchant trend in recent reflexivist fashion has been to trash previous generations of scholars, as exemplified in Geertz's transparent dismissal of Evans-Pritchard "Nuerosis" and, as the academic world turns, in Crapanzano's hermetic sealing and at times downright Bali-aching about "Who told Geertz?"[6] Indeed, there has been so much writing off of anthropologists and sociological proto-anthropologists who wrote about culture, and such a crescendo in critiquing those who carried a culture concept, that the following remarks may contain little of shock value. Especially, I should add, since Talal Asad has already offered a scathing critique of Gellner's "demonstrably faulty" approach to cultural translation and religion, Jon Anderson has deflated Gellner's conjuring of Islam, Henry Munson has shown that Gellner's segmentary model of Moroccan tribalism "did not exist," and Hugh Roberts demonstrates how Gellner misread earlier French ethnography on segmentation via a footnote-in-mouth mistake by Durkheim.[7] Gellner, himself, held little sacred in savaging those he disagreed with, including a fair number of Americans who had ventured into his sanctified field turf of the Moroccan Atlas. Yet, as Aristotle, Gellner's philosophical ancestor, once said: "Great men may make great mistakes."[8]

Gellner's *Fatiha*

The themes expounded in this book have been gestating for over a quarter of a century, ever since I first visited central Morocco, and my first debt

*was to those of its inhabitants who tolerated my intrusion. The central
ideas are plainly stolen from four great thinkers—Ibn Khaldun, David
Hume, Robert Montagne, and Edward Evans-Pritchard. The stream
that started in the central High Atlas with fieldwork experience was fed,
over the years, by many others—notably by systematic attention to the work
of other ethnographers working in the Muslim World. Much of that work
was only being produced during that period, and my next debt is to all
those anthropologists who shared their ideas and data with me, very often
prior to publication. (My attention to historical work was less systematic,
as no doubt will be evident to readers.) During much of this period, I took
part in running a seminar on the sociology of Islam.*[9]

Ernest Gellner

Opposite the title page of my paperback copy of Gellner's *Muslim
Society* are six accolades from published reviews—reminiscent of
the manner in which films are lauded by phrases interspersed
between. . . . The lead quote is from fellow Weber admirer, Clifford
Geertz, with whom Gellner is coeval and coequal in publishing a
book with only two words in the title. For Geertz, keeping reviewer's
license in mind, this is the "boldest and most ingenious . . . attempt
in recent years to present a general account of the fundamental
features of social life in the Islamic World."[10] Whether Geertz wished
to connote this boldness as a sense of daring or one of presumption,
this is an apt summation of the goal, and thereby the chief problem,
of Gellner's work. The British professor, befriending an ancient Arab
scholar who seemingly thought like a post-Enlightenment sociologist
and an Enlightenment philosopher of religion, is thus able to define
what is fundamental about the Islamic World by ignoring the writing
of all contemporary Muslims. Yet, as the fourth praiseworthy epigraph by Abbas Kelidar warns us, Gellner's conclusions are so reliant
on the politics of North Africa that "how applicable they are
elsewhere in the Muslim world remains a question." An important
question, indeed, for a book that entitles itself to "Muslim Society"
at large.

In another blurb, M. A. Zaki Badawi describes the book as "brilliant work" that should be of special interest to scholars "who uphold
the faith."[11] Serif Mardin styles this contribution as a series of "brilliant essays." It is, for Jacques Berque, "the kind of exposé that can
be read with genuine intellectual pleasure." Why? Because, as Mardin
observes, of its "sustained originality, compact argumentation and
pervasive wit." Finally, Ralf Dahrendorf avers that "Professor Ernest
Gellner is almost uniquely qualified to discuss the implications of this
development," that is, the revival of Islam. That Gellner is original,
witty, and eminently provocative as well as bitingly acerbic is readily

apparent to anyone reading his essays. That he is correct, or even on to the kind of insight that will revolutionize the field, is somehow missing from the conclusions in the carefully chosen quotes from his peers.

Book reviews are notoriously personal creations. A good review or a bad review often has less to do with the quality of the book than the prejudice—stated or otherwise—of the reviewer. How is it, then, that Professor Gellner had come to be regarded "almost uniquely qualified" to write such a book? A possible answer to this is provided in the rhetoric of Gellner's opening paragraph, quoted above. When a scholar has been writing about more or less the same themes for a quarter of a century, he can easily become—through no fault of his own—an expert. This is especially the case for a professor at a distinguished university, one who has long taken part in "running a seminar on the sociology of Islam" and who goes on to acknowledge an intellectual debt to a sample—"too numerous for exhaustive listing" of some thirty or more colleagues. Gellner's authority is self-presented as a scholar who follows on the heels of "great thinkers" and has paid "systematic attention to the work of other ethnographers working in the Muslim world."[12] And, the trump card for this British social anthropologist is that he was "there." Gellner's first debt, we are informed, is to the Moroccans who tolerated his ethnographic presence and thus now validate his ethnographic authority. Gellner, like Geertz, parades this mantle from the start.

Ironically, Gellner's attempts to defuse, and at times diffuse, criticism point to the major failings of his essays. This becomes a smokescreen maneuvering under the cloak of ethnographic authority, liberal acknowledgment of intellectual influences, an overwrought sense of debt to other ethnographers, a highly selective and severely constrained use of relevant historical materials, and an approach to Islam as something one runs seminars about. The choice of first person here allows the author to tell *you* in an informal manner what an expert *he* is without sounding unduly pompous. *I* went to Morocco, Gellner confides; *I* owe a debt to other great thinkers; *I* have paid attention to previous historical work. The reader is immediately disarmed by Gellner's carefree, somewhat self-deprecating, comments. The Moroccans "tolerated" his intrusion; his central ideas are "plainly stolen" and he owes an intellectual debt to many of the people with whom he has never been in a seminar. My criticism of *Muslim Society* thus takes as its starting point Gellner's telling admissions as clues to his untold omissions in concocting a highly idiosyncratic anthropology of Islam.

Saints of the Atlas, published in 1969, was the text that launched Gellner as an expert on Islam and Middle Eastern society. Expanding on the African model of segmentary tribal structure elaborated earlier by Evans-Pritchard for the Sudanese Nuer and Cyrenaican Bedouin, Gellner views the Berber society of mid-twentieth-century Morocco as a microcosm for the essence of Islam. As Gellner sets the geographical stage,

> Morocco prior to the twentieth century sounds like a parable on the human condition in general. The country could be seen as composed of three concentric circles: the Inner Circle of tribes who extracted taxes, the Middle Circle of tribes who had taxes extracted from them, and the Outer Circle of tribes who did not allow taxes to be extracted from them. In other words, there were the sheep-dogs, the sheep, and the wolves.[13]

Against this agonistic backdrop a succession of Islamic dynasties sought power by convincing some tribes to watch over the sheep—and the ruling shepherds—and also to keep the ungovernable wolves at bay. Foregoing analysis of formal Islamic intellectuals, Gellner suggests that the tribal lineages of holy men, known locally as *igguramen*, served as "Lords of the Marches" who played a dual role of pacifying the segmentary tribes for the government and protecting their spiritual clients from the government. Since the tribes were unable to govern themselves, the Berber *igguramen*, known in Arabic as marabouts, served as a focal point for political stability and religious legitimacy. The book provides an ethnographic portrait of the main lodge (*zawiya*) of Ahansal, founded at the end of the fourteenth century, and the relations between these local saints, the surrounding tribes, and the Moroccan government. In *Muslim Society* Gellner returns to his Moroccan example and Ibn Khaldun to propose a philosophical model of the history of Islam.

His choice of title for his ethnography says a lot about the intellectual context he is coming from. How "Orientalist" of an ethnographer to render the Berber term *igguramen*—already entrenched as a borrowed term in the extensive French literature on the subject—with the overtly Christian usage of "saint." Is this not of a piece with labeling African healers as "witch doctors" or presumed Nuer leaders as "leopard-skin chiefs"? It is certainly symptomatic of Gellner's doctoring of the native point of view through his own Europhilosophical terms. There is some irony in Gellner's ethnographic presence in the Atlas of Morocco. To carry hermeneutic license a bit too far, it might be imagined that the notion of "Atlas" in the classical sense of the Titan who held up the pillars of the universe, parallels Gellner's fundamentalizing of the pillars of Islam.

By his own admission, Gellner addresses Islam as a sociologist who talked with real Muslims. In his fieldwork these were primarily the "saints," with whom he lived "for months on end," the end limit here totaling "well over a year."[14] "In the course of my fieldwork in the central High Atlas of Morocco, I was struck by the firmness and emphasis of one principle which tribesmen invoked when discussing their own social organisation: it is absolutely essential for a man to have his place. . ." writes the anthropologist in his study.[15] So, one might expect that a significant, if not large, part of the information in this set of essays would derive in some observable way from this base of ethnographic data. But in *Muslim Society* all we have is Gellner's recollection of a conveniently unnamed tribal everyman in place of the specific informants. As Talal Asad complains, "indigenous discourses" of Muslims are "totally missing in Gellner's narrative."[16] Quite bluntly, he ignores real Moroccans as readily as Geertz does in *Islam Observed*. Indeed, one of the main stars of *Muslim Society* is a Tunisian historian who wrote sociologically, in Gellner's eyes, some six centuries earlier.

The Man and His *Muqaddima*

Thus, to sum up, we find that Ibn Khaldun flashed like a solitary star in a pervasive pall of darkness. He is the father of sociology, the inventor of the scientific method of human studies and the originator of the philosophy of history.[17]

Buddha Prakash

And is it not baffling that of the scores of hundreds of books and articles on Ibn Khaldun all but a score are worthless?[18]

Aziz Al Azmeh

So who is this individual that Gellner and so many scholars have over-idealized in print? He is—not in short—Wali al-Din 'Abd al-Rahman ibn Muhammad ibn Muhammad ibn Abi Bakr Muhammad ibn al-Hasan, known primarily as Ibn Khaldun.[19] Born in Tunis on May 27, 1332, his pedigree stretched back to a Yemeni Arab ancestor who settled in Seville during the early Muslim conquests. The relevant sources suggest that this was a well-regarded family with political connections in Tunis stemming back to Ibn Khaldun's great-grandfather, a financial vizier who ended up getting strangled to death after a local coup toppled the regime of the day. His father "wisely avoided politics," as the entry in the *Encyclopaedia of Islam* puts it, and made sure the young Abd al-Rahman received a thorough education. The dreaded Black Death that ravaged the middle of the fourteenth century left the future scholar an orphan at age seventeen. His life over the next two decades was a mixture of being a perpetual student and dodging political peril with administrative appointments worthy of a novelist's imagination. In later years he bounced around among cosmopolitan North African venues, from the intellectually stimulating Granada of Islamic Spain to resplendent Cairo under the Mamluks.[20] He served several political and legal roles, taught extensively and experienced a healthy quotient of adventures, capped by an interview with the Mongol warlord Tamerlane, who allegedly offered him a job.[21] We know many of the details of his life because Ibn Khaldun had the foresight to leave posterity an autobiography, not surprisingly about as self-servingly fleshed out as memoirs of flashing stars come.

The primary, if not the only, reason Gellner and other Western social scientists know about Ibn Khaldun is because of the long and intellectually fascinating introduction to his otherwise rather standard universal history. While Ibn Khaldun was apparently a prolific writer, the only major treatise that has survived is his general, known-world history with its eminently detachable introduction. The *Muqaddima* was written during some five months in 1377 C.E., while the author was in his mid-forties. As an introduction, there is little not covered from the sciences, arts, religion, language, history, and occult. Indeed "the most comprehensive synthesis in the Human Sciences ever achieved by the Arabs" is readily categorized by Charles Issawi into a table of contents guaranteed to whet the appetite of secular-minded Western scholars:[22]

Chapter One. Method

Chapter Two. Geography

Chapter Three. Economics

Chapter Four. Public Finance

Chapter Five. Population

Chapter Six. Society and State

Chapter Seven. Religion and Politics

Chapter Eight. Knowledge and Society

Chapter Nine. The Theory of Being and Theory of Knowledge.

Ibn Khaldun, as introduced through Issawi's eyes, escapes being medieval because "his positive outlook and matter-of-fact style render him particularly congenial to the modern mind, brought up on a tradition of scientific method."[23] The translator's goal is quite explicitly an attempt to "present Ibn Khaldun's thought in a style and terminology familiar to students of the social sciences and to avoid a literal translation which might obscure the depth and originality—one might say the modernity—of his theories." Indeed, one *should* say the modernity. In short, here is an Arab and a Muslim who might as well be Aristotle or Machiavelli, a free-floating, pre-Enlightenment free-thinker linking classical Greek philosophy to modernity.

As a self-promoting and politically judicious author, Ibn Khaldun was not shy about the novelty of his work. "It should be known that the discussion of this topic is something new, extraordinary, and highly useful," exclaims the savant. "Penetrating research has shown the way to it."[24] Although lost in a sense during the early rise of rationalist philosophy and modern social science, the *Muqaddima* has been wrung true to the author's prognosis.[25] Indeed, in the West it is hard to find any modern field of social science or history that Ibn Khaldun has not been cited as forerunner for. For example, Kamal Ayad scarcely misses an intellectual current in referring to the text as an "empirische-soziologische-biologische Geschichtsauffassung."[26] One need not read German fluently to catch the totalizing Achtung in this accolade. Ibn Khaldun becomes the erstwhile godfather of virtually every icon of Western intellectual tradition, including (in alphabetical order for the sake of not giving away my own forefatherly hierarchy): Bodin,[27] Collingwood,[28] Comte,[29] Cornet,[30] Darwin,[31] Durkheim,[32] Feuerbach,[33] Foucault,[34] Gumplowicz,[35] Hobbes,[36] Kant,[37] Sir Arthur Keith,[38] Machiavelli,[39] Montesquieu,[40] Nietzsche,[41] Pareto,[42] Adam Smith,[43] Herbert Spencer,[44] Spengler,[45] Spinoza,[46] Veblen,[47] Paul Valéry,[48] Vico,[49] and Weber.[50] He has also been dubbed an intellectual predecessor of lesser lights, at least sociologists whose visibility in the field has dimmed considerably, including Cooley, De Roberty, Draghicesco, and Izoulet.[51] Ibn Khaldun is such an entrenched sociological icon that James Davis, in an essay entitled "What's Wrong with Sociology?" believes that it was not until after the Vietnam War that Ibn Khaldun did not loom as large as Marx in the discipline.[52]

Ibn Khaldun is the Arab philosopher non-Muslims love to laud, but at what expense? The frontispiece to Issawi's translated excerpts provides a paradigmatic example of the passive–aggressive relationship Western intellectuals have had with this praiseworthy predecessor of their ideals. The reader is treated, as a pre-preface, to glowing tributes from canonical historians Arnold Toynbee, George Sarton, and Robert Flint.[53] Toynbee avers that Ibn Khaldun "conceived and formulated a philosophy of history which is undoubtedly the greatest work of its kind that has ever yet been created by any mind in any

time or place."[54] Here is an "Arab genius" on a par with Thucydides
and Machiavelli. But, Toynbee adds, Ibn Khaldun's star quality is all
the more remarkable because this one very forward-looking individual is "the one outstanding personality in the history of a civilization
whose social life on the whole was 'solitary, poor, brutish, and
short.'" It is hard to imagine a greater put-down of Islam that this
feigned tribute to one lone Muslim scholar, since the entire civilization is at the same time reduced to the savage level implied in
Toynbee's consciously chosen reference to Hobbes. That which
makes Ibn Khaldun shine is the almost total lack of any other rationally modern-looking individual in the millennium plus history since
the days of Muhammad. George Sarton likewise fixes Ibn Khaldun's
stardom as "the greatest historian of the Middle Ages" above all but
two medieval Christian scholars. How fittingly biased the mixed-
message metaphor that this lone Arab is "towering like a giant over a
tribe of pygmies." Robert Flint continues to illuminate Ibn Khaldun
as "a theorist on history" with no equal "in any age or country" until
Vico, and who orbits in a sphere above even Plato, Aristotle, and
Augustine! "He was, however, a man apart," suggests Flint, a scholar
"solitary and unique among his co-religionists . . ." The success of
Ibn Khaldun is thus the fact that he is so unlike every other Muslim
and Arab. As an adopted scion of Western thought, his brilliance only
serves to sharpen the dark hole of a civilization thought to be utterly
incapable of recognizing this one man's genius.

Readers of Ibn Khaldun's *Muqaddima* often imagine an invisible
hand inscribing the great truths of Western secular thought. Issawi
suggests that the Tunisian savant [pre]saged Locke and Hume and
was in effect "more clear-sighted than Adam Smith."[55] A Marxist–
Leninist scholar, S. M. Batseva, acknowledges this wily medieval courtier
as the first Marxist in history for recognizing "the role of labor as the
creator of value" and as a defender of the Maghribi masses.[56] In his
influential *Encyclopaedia of Islam* article, Talbi compares quotes by
Ibn Khaldun and Karl Marx on the economic basis of society. He then
cautions that "in spite of the undoubted similarities, it would be difficult to regard Ibn Khaldûn as a forerunner of materialism."[57]
Caveats aside, this is what much of the rhetoric does suggest. And
the Ibn Khaldun Economic Wizard trope continues. A contemporary
German economist quite recently hyperbolized in a respectable
journal:

> Ibn Khaldun rightly identified the relevant components of the process
> of economic development: creation of added value, the working mech-
> anism of supply and demand, consumption and production, the role of

money, capital formation and public finance, population growth, the
effects of urban agglomeration, the crucial role of agriculture,
the importance of political stability, and the conditions of the macro-
economic regulatory system as echoed in contemporary structural
adjustment programs.[58]

Apparently this author was unaware that he had been scooped in
1981 by Ronald Reagan, who called Ibn Khaldun the original author
of supply-side economics.[59]

Ernest Gellner is hardly any less economical in his praise of Ibn
Khaldun as a fourteenth-century economist who "elaborates a
Keynesian theory of economics."[60] Significantly, philosopher-turned-
Islamicist Gellner is keen on the "cold sociological eye" with which Ibn
Khaldun is said to view society and history.[61] More specifically, Ibn
Khaldun is a "positive, descriptive sociologist" rather than a "pre-
scriptive political philosopher," as though these are the only two
options Gellner thinks available.[62] He is not alone in seeing Ibn
Khaldun as the precursor of modern sociology; indeed, scholars were
calling this Arab scholar a sociologist while the late nineteenth-
century field of "modern" sociology was still taking form. Even the
noted German Orientalist Ignatz Goldziher acclaimed Ibn Khaldun
as the founder of sociology. Issawi lists six "basic principles on which
sociology must rest," all to be found either embryonic or full-flush in
the *Muqaddima*.[63] For Akbar Ahmed, Mahmoud Dhaouadi, and
Mohammed Abdullah Enan, Ibn Khaldun's social concept of *'umran
is* sociology.[64] Yet, as Jon Anderson rightly observes, it is better to
regard this fourteenth-century Tunisian intellectual as a "kindred
spirit" rather than "a totemic ancestor."[65]

To be fair, many of the Western scholars who praise Ibn Khaldun
tend to be those who never read him in the original Arabic. Probably
no scholar is better versed in the intricacies of Ibn Khaldun's prose
than the superb Arabist Franz Rosenthal, who translated the complete
Muqaddima into English over four decades ago. After observing that
there seemed to be hardly any great thinker with whom Ibn Khaldun
had not been compared, Rosenthal cautions: "Such comparisons may
help to evaluate the intellectual stature of the person with whom Ibn
Khaldûn is compared; certainly they suggest a lesson in scholarly
humility. But they do not contribute much to our understanding of
Ibn Khaldûn."[66] As early as 1933 the historian H. A. R. Gibb had
remarked that much of Ibn Khaldun's thought has a moral and religious basis rather than a sociological one.[67] Gibb further observed
that "the axioms or principles on which his study rests are those of
practically all the earlier Sunni jurists and social philosophers."[68]

Another historian of Islam, Gustave von Grunebaum, further deflated Ibn Khaldun's originality by claiming that the same issues had been dealt with by al-Mas'udi, four centuries earlier than the era of the *Muqaddima*.[69] Much of this misuse of Ibn Khaldun's ideas could have been avoiding by paying attention to the consensus of Arabists that he was not necessarily the genius he is invariably touted to be. David Dunlop, in an influential survey, concluded: "Most of the faults for which he blamed his predecessors are conspicuous in his own *History*. Nor could it be otherwise, since for the most part he contents himself with abridging them."[70]

How savvy was the proposed Pater Arabicus of Sociology? There is no dispute that the *Muqaddima* is one of the most wide-ranging and suggestive historiographic texts ever written in Arabic. The problem arises when a late fourteenth-century author is teleologicized out of his time and space, then repositioned in a "process in which he certainly did not participate, but to which his attachment is made imperative because of the apparently dis-Islamic character of his thought."[71] Numerous Western scholars have found Ibn Khaldun to share an *esprit d'corps* with their own secular, positivist agenda. Consider the hyperbolic praise of Brunschvig:

> Just as he had no forerunners among Arabic writers, so he had no successors or emulators in this idiom until the contemporary period. Although he had a certain influence in Egypt on some writers of the end of the Middle Ages, it can be stated that, in his native Barbary, neither his *Mukaddima* nor his personal teaching left any permanent mark. And indeed systematic lack of comprehension and the resolute hostility which this nonconformist thinker of genius encountered among his own people forms one of the most moving dramas, one of the saddest and most significant pages in the history of Muslim culture.[72]

It is thus alright to admit this "Oriental" into the intellectual club because he is so unlike all those other Arabs and Muslims; he might as well be European, except for the fact that most European scholars at the time were not as modern in historical hindsight. There is a tendency in the excessive literature on Ibn Khaldun to read the cultural commentary and ignore the theology. Here, it seems, is one Muslim scholar for whom church and state can be readily separated.

"In short," argues Bruce Lawrence, "Ibn Khaldun is a product of Orientalism . . ."[73] Thus it should come as no surprise that Western adoration of Ibn Khaldun as an odd-man-out of Arab scholarship generated a vigorous backlash from Arab and Muslim scholars. Antoine Makdisi, for example, writes indignantly against the Western scholarly beatification of Ibn Khaldun as though Arab greatness must be measured by what the modern West considers great.[74] This rejection actually crosses the range of religious perspectives, since institutions like al-Azhar have long been the scene of skepticism about the value of the *Muqaddima* for promoting Islamic knowledge. Progressives like Muhammad Abduh might have gravitated toward Ibn Khaldun, but Rashid Rida and Taha Hussein were turned off by exactly what turned on Western commentators.[75] Yet despite the earlier countercriticism, Ibn Khaldun remains for many Arabs a bellwether on what defines a scholar.[76]

Of particular interest to Gellner and many others is Ibn Khaldun's cyclical model of dynastic half-life.[77] This has been described, redescribed, and oversubscribed for over a century and a half. In a nutshell, yet again, the eternal opposition between the nomads (*'umran badawi*) and the settled folk (*'umran hadari*) sets the stage for state creation, stagnation, and renewal. "The first stage is that of success," states Ibn Khaldun, "the overthrow of all opposition, and the appropriation of royal authority from the preceding dynasty." The new hero ruler comes to power by virtue of a strong bond of group feeling, called *'asabiyya* by Ibn Khaldun. In the next stage the ruler takes over exclusive control, consolidating his power by distancing from his equals and relatives, thus relying on strangers and mercenaries. Here blossoms the golden age of state projects, financed by heavy taxes, and fancier uniforms for the military. The fourth stage is one of imitation, resting on past laurels, refusing to think in new ways. All this leads to the fifth and culminating part of the dynastic cycle: "waste and squandering." Bad and low-class lackeys take over, the military grumbles and "senility" and "chronic disease" set in. The stage is thus set for some new desert Bedouin, bursting with group feeling, to overthrow the old and start it all over again.

It is understandable why later scholars would latch onto this particular explanation of dynastic change in North Africa, since it appears to offer a sociopolitical argument rather than attributing change to the will of Allah. The mechanism that most sticks out in post-Enlightenmentality is the unique and novel concept of *'asabiyya*. This term, which appears more than 500 times in the *Muqaddima*, has been variously translated as "social solidarity,"[78] "social cohesion,"[79] "tribal spirit,"[80] Gemeinsinn,[81] *esprit d'corps*,[82] *spirito di corpo*,[83] "vitality,"[84] and even a racial philosophy.[85] Fischel describes it as "the most powerful force in the creation and development, rise, duration and fall of a religion, society or nation."[86] Circling back to Western tradition, Helmut Ritter equates *'asabiyya* with Machiavelli's power-driven *virtù*.[87] Whatever Ibn Khaldun meant at the time, an enigma

not likely to be resolved by Western scholars alone, the sentiment of
Yves Lacoste is worth repeating: "Virtually everyone who has written
on Ibn Khaldun has his own interpretation of *'asabiya*."[88]

In an early précis of Ibn Khaldun, Gellner reads over this sociologist's dream model in Aesopian delight as a North African variant of
"survival of the fittest" in which the wolves come out of the wild
desert to replace the exhausted, urbane sheepdogs.[89] Although the
fourteenth-century author's prose is quite clear, Gellner adds an
interpretive layer of two contrasting and competing religious
syndromes: one urban, one tribal. For Gellner the urban variety,
arbitrarily labeled "p," features:

> Stress on Scripture and hence on literacy. Puritanism, absence of graven
> images. Strict monotheism. Egalitarianism as between believers;
> minimisation of hierarcy. Absence of mediation, abstention from ritual
> excesses. Correspondingly, a tendency towards moderation and sobri-
> ety. A stress on the observance of rules rather than on emotional states.

This is the world guarded, ultimately in vain, by the sheepdogs. Then
there is the "c" syndrome for the rural tribes:

> Personalisation of religion, tendency to anthropolatry. Ritual indul-
> gence, and absence of puritanism. Proliferation of the sacred, concrete
> images of it. Religious pluralism in this and the other world and local
> incarnation of the sacred. Hierarchy and mediation.

For the animal metaphor lover, this is the religion of the wolves.
Sheepdog vs. wolf, town vs. tribe, doctor vs. saint; such is the binary
interplay defining the "characteristic Muslim state" as a "distinct and
characteristic type of its own."[90]

The problem with Gellner's idealization is that it fits neither the
description of dynastic change given by Ibn Khaldun nor the historical and ethnographic evidence for North Africa. The illogic in the
scheme is staggering. The urban devout are egalitarian and the tribes
are hierarchical! Townsmen are fanatical monotheists and tribesmen
are pluralistic self-worshippers![91] Gellner's odd pairings are more
easily understood by his own admission that all this has less to do with
Ibn Khaldun than the simple fact that North African Islam "holds up
a mirror-image to Western Europe."[92] The rational for his pc model
unravels as politically incorrect, unless it is innocent coincidence that
in Europe the "c" syndrome stands for centralized Catholicism and
the "p" for the "minoritarian, fragmented and discontinuous"
Protestants; Gellner's model would in fact invert such a correlation.

"The contrast is striking," as Gellner feels compelled to interject, but it is also hopelessly flawed as a model to understand Muslim society.[93]

Gellner, however, is undaunted in his attempt to mine the *Muqaddima* for modern-looking anthropological theory. "Long before modern social anthropology made the same discovery, Ibn Khaldun knew full well that the state of nature is not individualistic, but tribal."[94] His initial dependent clause is chronologically on target, but the primary point is self-incriminating. First, Ibn Khaldun knew, like nearly everyone else in his time, that man was essentially a product of his environment. "The camels are the cause of (the Arabs') savage life in the desert," the Arab scholar notes.[95] Other climes—there were thought to be seven in all—other environmental determinants. In his day Ibn Khaldun also knew "that most of the Negroes of the first zone dwell in caves and thickets, eat herbs, live in savage isolation and do not congregate, and eat each other."[96] Some modern-day sociologist! To add yet another allegedly anthropological insight, consider his spin on kinship: that lineages remain pure in the desert because no civilized person wanted anything to do with them.[97] Some habitus! Second, Gellner takes pride in a position that most anthropologists on either side of the Atlantic have now abandoned. The segmentary lineage model that Gellner rides has no leg to stand on and has been mercifully put out of its misery. Certainly the state of nature, including culture, is not individualistic; the available options go far beyond an idealized tribal model that no one ever actually observes in operation.

Flux and Redux: [Ex]Huming Islam

The model which is here offered of traditional Muslim civilisation is basically an attempt to fuse Ibn Khaldun's political sociology with David Hume's oscillation theory of religion.[98]

Ernest Gellner

It is provocative, though highly idiosyncratic, to argue that the "best approach to the social role of Islam is probably through the religious sociology of David Hume," a philosopher who Gellner admits "is not normally considered an Islamicist."[99] It is equally the case that Hume's enlightened and natural skepticism, shared by modern-day scholars such as Gellner himself, does not transform him into a latter-day "sociologist" anymore than it does pamphleteer Thomas Paine.[100] If we take Hume at his word that his concern with religion was in relation to its "foundation in reason" and "origin in human nature," we have a philosophical enterprise—indeed, a highly influential one—but not social theory. I am indeed at a loss to see in what respect Hume provides the first "scientific study of the place of religion in society," unless for Gellner scientific only means a certain British style of skepticism.[101] Hume nowhere "observes" religion as a social phenomenon apart from the limited historical examples available at armchair length in his library. When he remarks that the "savage tribes of AMERICA, AFRICA, and ASIA are all idolaters,"[102] are we to assume that he is scientifically rendering the wild travel accounts he consulted? The Scottish philosopher surveys the history he knows, but his ultimate claim is that religion, particularly the monotheistic variant, pales next to a rationally derived moral philosophy.[103] Hume is a significant and pathbreaking philosopher of religion; why sully his deserved reputation by forcing him into the mold of a proto-sociologist, as, ironically, Gellner is also prone to do for Ibn Khaldun?

What Gellner discovers in Hume is a theory that previous anthropologists and philosophers somehow failed to notice; indeed, we are assured that the modern-day social theorist has exhumed what is "most distinctive and important in Hume's sociology of religion." This long dormant theory, now brought out to best explain Islam, is a "central, interesting and profound oscillation theory."[104] The crux of this theoretical redux is flux and reflux. As quoted by Gellner, Hume had argued: "It is remarkable that the principles of religion have a kind of flux and reflux in the human mind, and that men have a natural tendency to rise from idolatry to theism, and to sink again from theism to idolatry."[105] Thus, Gellner informs us, "the heart of Hume's theory of religion" is that men change from polytheism to monotheism and back again "not for rational reasons" but out of fear. Gellner's reading here is rhetorically flawed and most unreasonable, playing on the ambiguity of "reason" after more than two centuries of academic overuse.

Counter exegesis can best be made by placing Hume's "theory" back into its textual context:

> It is remarkable, that the principles of religion have a kind of flux and reflux in the human mind, and that men have a natural tendency to rise from idolatry to theism, and to sink again from theism to idolatry. The vulgar, that is, indeed, all mankind, a few excepted, being ignorant and uninstructed, never elevate their contemplation to the heavens, or penetrate by their disquisitions into the secret structure of vegetable or animal bodies; so far as to discover a supreme mind or original providence, which bestowed order on every part of nature. They consider these admirable works in a more confined and selfish view; and finding their own happiness and misery to depend on the secret influence and

unforeseen concurrence of external objects, they regard, with perpetual attention, the *unknown causes*, which govern all these natural events, and distribute pleasure and pain, good and ill, by their powerful, but silent, operation. The unknown causes are still appealed to on every emergence; and in this general appearance or confused image, are the perpetual objects of human hopes and fears, wishes and apprehensions. By degrees, the active imagination of men, uneasy in this abstract conception of objects, about which it is incessantly employed, begins to render them more particular, and to clothe them in shapes more suitable to its natural comprehension. It represents them to be sensible, intelligent beings, like mankind; actuated by love and hatred, and flexible by gifts and entreaties, by prayers and sacrifices. Hence the origin of religion: And hence the origin of idolatry or polytheism.[106]

Here is the flux *prima facie*, a secular claim-staking that would be taken up and expanded later by sociologists and anthropologists.[107] Hume's point, unlike the standard line with conservative theists of his day, is that religion or a sense of deity is in fact understandable through reason as a natural act rather than falling back on revelation as a dogmatic belief. Perhaps responding to the psalmist who preaches, "The heavens declare the glory of God" (Psalm 19:1), Hume rationalizes that the "ignorant and uninstructed" nature of earliest humanity would not have been capable of recognizing this assumed glory. In such a state "unknown causes" would be attributed to "pleasure and pain, good and ill." Then the "active imagination" of men would anthropomorphize such causes into "sensible, intelligent beings," thus accounting for the worship of spirits or gods, that is, idolatry in the biblical sense. Indeed, for Hume even the most sublime Christian and Muslim theologians were essentially pursuing a philosophical "anthropology" that was anthropomorphic to a fault.

The reflux follows from a similar natural tendency in human nature as "exaggerated praises and compliments" elevate the deities to the monotheistic markers of "unity and infinity, simplicity and spirituality." But once such a "lofty" state is achieved, the banished idolatry inevitably returns. Indeed, argues Hume, "so great is the propensity, in this alternate revolution of human sentiments, to return to idolatry, that the utmost precaution is not able effectually to prevent it."[108] It is for this very reason, continues Hume, that Jews and Muslims prohibited representations of human figures. Thus the oscillation, which is not a term used by Hume, is between "opposite sentiments" of not being able intellectually to conceive of "a pure spirit and perfect intelligence" and yet at the same time fear that their deities would have "limitation and imperfection." The whole psychological argument revolves around fear and anxiety; "The primary religion of mankind," theorizes Hume, "arises chiefly from an anxious fear of future events . . ."[109] It is thus Hume's reasonable assumption that as panic seizes the mind, "the active fancy still farther multiplies the objects of terror" and eventually this comes into conflict with an equally natural "propensity to adulation" in humans. The root of the matter, for Hume, is that religion "springs from the essential and universal properties of human nature" rather than having been breathed god-given in a mythical garden.[110] Humean human nature, speculated upon before a modern sense of human evolution, genetics or neurology, is a philosophical black box, which Gellner cleverly turns into a theoretically open-ended Pandora's box.

It is instructive to look at the way in which Gellner frames Hume over the several pages in which he appears to be letting Hume's own words elaborate this formidable theory that no one had fully appreciated before. The reader might reasonably—and wrongly, in this case—assume that the seventeen or so quotes Gellner garners from Hume are sequential. By not providing the exact citations in footnotes, although there are ninety-nine notes to the first essay, Gellner is able to hypostasize from Hume's prose as he pleases. But why is it that several generations of Hume commentators had "seldom noticed" the oscillation theory in this posthum[e]ous sense? Rather than broach this cross-disciplinal breach of etiquette, Gellner proves his point to his own satisfaction by simply chiding several earlier scholars for not seeing the theory that "preoccupied" Hume and is "central to his argument." First on the list is his anthropological colleague, E. E. Evans-Pritchard, whose *Theories of Primitive Religion* (1965) is something of a milestone in the intellectual prehistory of an explicitly anthropological approach to religion. Evans-Pritchard, as Gellner notes, brought up Hume as an example of someone who simply thought polytheism or idolatry preceded monotheism, a raging issue in what Evans-Pritchard rightly labels "pre-anthropological times."[111] Unlike Gellner, Evans-Pritchard pointedly does not treat Hume as espousing a sociological theory of religion. Perhaps Gellner should have heeded the advice of another anthropological colleague, Rodney Needham, who cautioned that "no philosopher will need to be told again, and by me, what were the characteristic views of Hume . . ."[112]

Having found the anthropologists wanting, the philosophers are equally scolded for being blind to the oscillation theory; in this case a single philosopher in one conference proceedings is the sacrificial goat for this unoriginal synecdochalism of the field Hume helped shape. The sin of Bernard Williams, the philosopher in question, is a sin of omission: he did not discuss the oscillation theory.[113] Gellner's sin, on the other hand, is one of commission, contending in a footnote that Hume did not assume the thesis that polytheism precedes monotheism. If so, this should certainly come as a surprise to anyone who reads the title of Hume's first chapter "That Polytheism was the Primary Religion of Man."[114] The flaw in Gellnerian logic here is a [sin]tactical bait and switch. Hume was not willing, it seems, to accept a unilineal evolutionary up-from-the-idol history of religion, but rather suggested a cyclical shift from polytheism to monotheism and back again to polytheism. Nowhere does Hume argue that monotheism would have been primary, because "theism" as such is secondary to human experience as such.

Having laid out this oscillating scenario by reordering the natural flow of Hume's quotes, Gellner goes on to note the obvious weaknesses with such a view: "it is profoundly psychologistic, locating the mechanism of the pendulum-swing in the human heart, and rather neglecting the society within which the changes occur; and it also contains a profound contradiction."[115] "Neglecting" society does not, at least for Gellner's vision of the craft, prevent Hume from offering the best approach to the social role of Islam! "Flux and reflux" for Hume, however Gellner chooses to dress it up as theory, is sentimental, an interplay of "feeble apprehensions" and "natural terrors." Gellner recognizes several critical problems with Hume's statements, but informs the reader that the oscillation theory "only needs to be refined and elaborated."[116] Gellner has quite a bit of refining to do, since Hume is accused of holding observations in one work that "are in blatant conflict" with the brilliant insights in another. So refined, Hume "comes close to formulating a Protestant Ethic theory" about the rise of "modern liberal society." To insist that Hume is concerned with the social in religion is a glaring contradiction; to translate the philosopher Hume into a forerunner of the sociologist Weber borders on the perverse.[117]

Perversion is an apt term to characterize the very choice of the arch-advocate of human reason unaided by divine revelation, David Hume, as a fitting interpreter of Islam. Mahometanism, the term used by Hume, was dubbed an inconsistent religion that "sometimes painted the Deity in the most sublime colours, as the creator of heaven and earth; sometimes degraded him nearly to the level with human creatures in his powers and faculties; while at the same time it ascribed to him suitable infirmities, passions, and partialities, of the moral kind . . ."[118] More bluntly, in a passage that well illustrates Hume's own frustration with the "intolerance of almost all religions," is how "Mahometanism set out with still more [more than the Jews] bloody principles; and even to this day, deals out damnation, though not fire and faggot, to all other sects."[119] Indeed, Islam appears in *The Natural History of Religion* primarily as a foil to attack Catholicism, most notably in the following anecdote:

A famous general, at that time in the MUSCOVITE service, having come to PARIS for the recovery of his wounds, brought along with him a young TURK, whom he had taken prisoner. Some of the doctors of the SORBONNE (who are altogether as positive as the dervishes of CONSTANTINOPLE) thinking it a pity, that the poor TURK should be damned for want of instruction, solicited MUSTAPHA very hard to turn Christian, and promised him, for his encouragement, plenty of good wine in this world, and paradise in the next. These allurements were too powerful to be resisted; and therefore, having been well instructed and catechized, he at last agreed to receive the sacraments of baptism and the Lord's supper. The priest, however, to make every thing sure and solid, still continued his instructions, and began the next day with the usual question, *How many Gods are there? None at all,* replies BENEDICT; for that was his new name. *How! None at all!* cries the priest. *To be sure,* said the honest proselyte. *You have told me all along that there is but one God: and yesterday I eat him.*[120]

Such is the relevance of Hume for understanding the faith of Islam: a silly anecdote that can only yield an oscillating asymptote. Or, as Hume himself might have responded, "Generally speaking, the errors in religion are dangerous; those in philosophy only ridiculous."[121] I think Hume would include Gellner's sociology under philosophical errors.

Facile and False: The Gellnerian Spin on Muslim Society

This is facile and false.[122]
Ernest Gellner

In his tacking back and forth between Ibn Khaldun and contemporary ethnography, Gellner's analytical point of reference is neither.[123]
Jon Anderson

Oscillations aside, it is instructive to look at how Gellner frames his modeling of Islam in the first and main chapter of *Muslim Society.* Very much the pragmatist, the author is blunt from the first sentence: "Islam is the blueprint of a social order."[124] Rather than representing such a blueprint in the architecture of the religion itself, citing Quran or Muslim clerics for example, we first learn about Islam through the eyes of an odd traveling theorist, Alexis de Tocqueville. Why should a nineteenth-century Frenchman most known for his penetrating analysis of the emerging American psyche be the initial expert, especially since the quote provided seems to require an ellipsis?[125] The point being made is that Islam has no "church" and no "priesthood," hardly a novel discovery for any European writer. Such comforting anglicized terms—certainly not as jarring as *sunna, hadith or shariah*—lead up to what appears to be the most important point of comparison: "Judaism and Christianity are also blueprints of a social order, but rather less so than Islam." The social make-up—pun fully intended—of Islam thus skyscrapers over its older monotheistic siblings. Those who read on are shortly informed that Shi'ism, the minority of the two major sects, is "closer to Christianity." Gellner approaches Muslim society primarily for how it differs from Christian society in the West. *Muslim Society* treats society as if Islam itself, as Muslims define it, does not matter.

The tropic delight of wordsmith Gellner is understanding Islam *vis-à-vis* the two faiths assumed to be far more familiar to the reader. Perhaps he is following the pattern of Evans-Pritchard's analysis of Nuer belief in relation to the Old Testament notions of Jehovah. If, of course, there is no separation of "church" and "state," then it would be absurd to follow the Gospel advice to "render unto Caesar" what belongs to Caesar. Ironically, the term "Caesar" appears five times on the first two pages of *Muslim Society*, while the name Muhammad does not surface until the sixth page; and that is in a quote from historian Marshall Hodgson! Appearing nominally before the prophet of Islam in a book about Islam are the historians Michael Cook and Edward Gibbon, folklorist extraordinaire Sir James Frazer, philosopher David Hume, church father St. Augustine, sociologist Max Weber, psychologist C. G. Jung, author T. S. Eliot, and the "Great Mother . . . [goddess]." Nor will the reader find the usually ubiquitous "five pillars" of Islam constructed from the blueprint plans.

Before reaching, or shall I say overreaching, Hume, the contemporary British scholar imagines "what would have happened had the Arabs won at Poitiers and gone on to conquer and Islamicize Europe."[126] Here is Gellner at his witticism best. In such a scenario he suggests we would now be reading Ibn Weber's *The Kharejite Ethic and the Spirit of Capitalism* "which would conclusively demonstrate how the modern rational spirit and its expression in business and bureaucratic organisation could only have arisen in consequence of the sixteenth-century neo-Kharejite puritanism in northern Europe."[127] We are also assured that a Muslim Hegel would have avoided the "embarrassing boob" of claiming the earlier faith of

Christianity was "more final and absolute" than the later Islam.[128] All of this might very well have led to a more satisfactory view of history in which "Islam is, of the three great Western monotheisms, the one closest to modernity." A full week of pages into the book and Gellner has only managed to show how Islam is not Christianity and could have been a friend of modern secularism.

The issue for Gellner is not doctrine nor what Muslims believe differently from Christians; this indeed had been the primary interest in Islam for David Hume. Muslim society is represented solely as a politicized arena in which Islam is its own Caesar. "Islam officially has no 'church,'" argues Gellner, intentionally italicizing the ambiguous use of "church" here.[129] By harking back to the Western dichotomy between "church and state," this would imply that Islam is all "state." But the reader of sociologist Gellner should also be aware of Durkheim's insistence that each religion has a "church" in the sense of a formalized institutional structure.[130] The Islamic *ummah*, caliphs, imams and even "saints" constitute parts of the Islamic "church" in a Durkheimian model. If all Gellner wishes to say is that Islam has no "church" in the same way that Christianity does, this is hardly a novel idea. It is related to the contemporary Bernard Lewis-ism that Islam has not yet seen the enlightenment of secular democracy. Gellner is playing rhetorically here with the fire of religious ideology on both ends. He would argue, as a non-Muslim, that there is no "overall organisation, which could thereafter monopolise and rationalise sanctity and magic."[131] But certainly a great many Muslims think such a community is exactly what the ideal *ummah* should be. So is "Muslim society" what a historian might say it has been in specific cultural contexts or what a devout Muslim believes it should be?

Christianity and Judaism are intellectually detached by Gellner from the dirty politics of empire building. "Christianity, which initially flourished among the politically disinherited, did not then presume to *be* Caesar," contends Gellner. This is an odd spin on history for a nontheologian. Does Gellner base this claim on the gospel truth of a Christian *hadith* in the New Testament book of Acts, and has he taken the time to read up on the historical evidence for the early church? Certainly by the time of Constantine, the "church" became "Caesar." Did Henry the VIII create the Anglican faith in order to separate church and state as a principle or to legitimize his power through religion? To claim that Christianity, outside specific historical contexts, has "a kind of potential for political modesty" with only intermittent "theocratic aspirations," as Gellner does unabashedly, would require rewriting most of European history. Perhaps Gellner confuses the lack of a dominant worldwide Christian theocracy with the attempts over and over again over centuries to build political power through Christian religious rhetoric. Thus Islam is portrayed to the reader as a unique melding of politics and religious ideology by ignoring how true this has also been in the evolution of Christian Europe and by assuming that there were no viable indigenous discourses of political resistance in Muslim societies. I cannot but agree with Talal Asad: "As an anthropologist, however, I find it impossible to accept that Christian practice and discourse throughout history have been less intimately concerned with the uses of political power for religious purposes than the practice and discourse of Muslims."[132] Gellner's spin on the lack of an Islamic "church" is misstated.

A central thrust of Gellner's argument is that Islam has adapted to nationalism as a political strategy in which Muslim reformists can bypass secularism as seen in the West, and maintain a sense of religious purpose. Two models of Muslim modernization are proposed: with and against religion. Kemalism in modern Turkey exemplifies the first in which "Islam-shackled-to-the-state" leads to "an at least relative secularism."[133] Yet the origin of Kemalism is dismissed as "secularism in an unwittingly Koranic, puritanical and uncompromising spirit."[134] At the time Gellner saw the other option of a "Reformed Islam" as a viable alternative for sweeping away rural superstition and at the same assuming a "mantle of orthodoxy" that does not necessarily threaten the nation-state. He recognized the unsettling influence of "fundamentalism," including the Khomeini revolution in Iran, but still was convinced that it could be "socially and intellectually attractive to separate a true, pristine, pure faith from the superstitious accretions" and that this could be done "with real conviction."[135] Recent critics, admittedly with the advantage of fertile hindsight, have found flaws in Gellner's analysis. Robert Hefner, for example, observes that Gellner "exaggerates the degree to which the non-Muslim world has been secularized" and thus fails to see that "Islam" is not a unique case.[136]

Islam mediated via David Hume and Ibn Khaldun drives the narrative until Gellner returns, halfway through the first chapter, to his idiosyncratic "segmentary" model of Muslim society. "Segmentary theory explains the cohesion and co-operation of groups, notwithstanding the fact that they are devoid of strong leadership or effective central institutions . . ." argues the field ethnographer back home in philosophic mode.[137] Defending a reified kinship ideal to the hilt, the whole history of Islam is thus synecdoche-yoked to a tribal metaphor very like what Gellner, virtually alone, thinks he saw in the history of Morocco. But there is a problem here. "The Maghrebin data which initially inspired the general model may seem an exiguous base for such ambitious claims," Gellner freely but disingenuously admits.[138] A question must therefore be begged. "Is the segmentary model a good account of the internal organisation of Muslim tribal societies?"[139] Many anthropologists who have worked in Morocco after Gellner would say "no." Henry Munson, for example, rethinks Gellner's description of Moroccan tribal structure to the point of categorically rejecting it.[140] Talal Asad, who studied pastoral nomads in Sudan, likewise rejects Gellner's fixation on a specifically "tribal" form of Islam.[141] But even if Gellner has correctly animated the history of Morocco as a back-and-forth battle between wolves and sheepdogs, the main problem is that an ahistorical trajectory is assumed by Gellner for the origins of Islam. The Islam-founding Arabs appear only as segmented Bedouins for the British anthropologist; perhaps he takes Ibn Khaldun's *'asabiyya* too assiduously?

Bypassing the merits of the argument as an anthropological issue, consider the following flawed rhetoric in a passage that serves as a prime example of Gellnerian doublespeak:

> It is pastoralism which inclines, without forcing, societies towards segmentary organisation, which then spreads towards adjoining agrarian communities by a kind of osmosis, impregnating them with the pastoral ethos and obliging them to emulate a form of organisation which helps them defend themselves.[142]

Once upon a time there were solidarity-conscious nomads who rode "me and my brother against my cousin" dynamics to create an idyll no self-respecting farming community could do without. Societies would only be inclined, not forced, to emulate their desert brethren, of course. Harking back to Ibn Khaldun, this would be the incline before the decline, a slippery slope indeed. The key methodological notion here is "a kind of osmosis," not even straight osmosis. Inclined or not, the settled folk could not avoid being obligingly impregnated by an "ethos," pastoral or otherwise. In Hobbesian homiletics—and Gellner is excessively fond of quoting Hobbes—it is distinctively advantageous for villagers to defend themselves as would wild and acephalous tribesmen rather than develop a bureaucracy or look to supreme rulers for political deliverance. But Muhammad was no nomad. The message of Islam evolved in the trade towns of Mecca and Medina; Islamic law and theology were codified in cosmopolitan urban centers like Damascus, Baghdad, and Cairo. The flux and reflux of Gellner's idealized Islam is thus as barren as his metaphor.[143]

So as not to leave this critique in a state of flux, a final reflexivist revisit to the rhetoric of Gellner's representation of Islam is relevant.

The problems with *Muslim Society*, certainly the more original first chapter, revolve around a failure by the author to leave his own intellectual perch and consider the diverse ways in which Muslims represent themselves. This book is written for a Western audience steeped in Enlightenment philosophy and steepled in Christian theology.[144] Allah help the stray Muslim scholar who heeds the call of M. A. Zaki Badawi, as noted on the frontispiece, to read this "brilliant work." Like Geertz, yet in many ways not quite like Geertz, Gellner never backs down from an intellectualized and essentialized "monolithic conception of Islam."[145] We never really learn what Islam means in praxis, certainly for ordinary Muslims, but instead are regaled with ideals rephrased out of original context. Along the way, the musings of Western heroes like de Tocqueville and Hume, neither of whom had any expertise on Islam, become definitive by default. And, as Vincent Cornell shows, Gellner's "after-the-fact explanations do not extricate him from the mire of reductionistic and tautological definitions."[146]

How can Gellner spin Islam in so facile and false a manner and manage to get away with it? I suggest there are three main factors to represent his representation. First, he enters the arena with the authority of an expert who has lived among Muslims cheek by jowl, sheik by *igguramen*. He is certainly not a historian or Arabist, nor does he ever need to claim to be. Here is the anthropologist comfortable in villages, not the Orientalist reading texts. Second, his style of disputation is full of quotes and references to many of the great intellectuals, at least pre-postmodern, in Western tradition. Sheer citational momentum thus validates his intellectual acumen. Third, Gellner is a master of rhetorical persuasion; his way with words sells without having to prove his point with detailed ethnographic documentation. Thus, *Muslim Society* mirrors *Islam Observed* in relegating actual Muslims to the theoretical sidelines. That Professor Gellner long held prestigious posts at major British universities certainly must be added into the equation, but such academic status is not a guaranteed trump card in the give-and-take-but-mostly-give maelstrom of British intellectual debate.

Ethnography is the missing ingredient, an oversight most non-anthropologists might actually find refreshing. Even when real Muslims are referred to, they "do not speak, they do not think, they *behave*" like dramatic actors following a script.[147] But Gellner stakes his claim to expertise over, and perhaps above, Islam only on the basis of fieldwork in the High Atlas of Morocco. His *Saints of the Atlas* focuses on the political organization—the fading passion of British social functionalism at the time—of tribes and their relationship to "saints," Gellner's popular rendering of both Arabic "marabout" and Berber "igurramen."[148] *Muslim Society*, parts of which were written as early as 1968, revisits a number of the themes first raised in the main ethnography. Oddly, Gellner does not specify for the reader the exact amount of time spent in Morocco. "I worked under a variety of opposed administrations, a veritable Vicar of Bray": acknowledges Gellner, "a French military one, a Moroccan Leftist one, a Moroccan military one, and finally, a Moroccan royalist civilian one again."[149] His informants, the *corpus delicti* of ethnographic authority, were "too numerous to list," although several who were "particularly helpful" do get a mention.

There is no doubt that Gellner conducted ethnographic fieldwork, nor do I wish to dispute the accuracy of his observations; but note how the reader of *Saints of the Atlas* is persuaded of the author's credentials as a generic expert from the start. Gellner was "there" through thick and thin. The field-graduated expert can speak about any aspect of Islam because he studied certain tribal and religiously charismatic Arabs and Berbers in a few locales of Morocco. Furthermore, Gellner would have us believe that he was there early on, when things were properly tribal and not tainted by recent cultural change. In responding to his critics, Gellner retorts, "Scholars who take the 'illusion' view of the segmentary egalitarian idiom tend to have done their field research fairly late in the development of these societies, I suspect they mistake what is indeed a correct account of the present, for one which was also valid in the past."[150] For the record, *all* ethnographic research in the region has been "fairly late," unless the field is expanded to ethnologists and folklorists of the early French colonial era. Gellner conveniently ignores the work of an earlier ethnographer, Emyrs Peters, who had come to doubt the power of the segmentary model.[151]

One potential measure of a scholar's work is an inventory of other men quoted. In Gellner's case there appear to be few women worth quoting. Such a list for Gellner's "flux and reflux" essay produces the following results, disaggregated by categories that reflect the arbitrariness of disciplinary boundaries in any kind of diachronic sequence:[152]

Sociologists/Anthropologists
Jacques Berque
J. P. Charnay
Sir Edward Evans-Pritchard
Emile Durkheim
Sir James Frazer

Milton Friedman[153]
R. Gallisot and G. Badia[154]
Dr. Riaz Hassan
Raymond Jamous
Ibn Khaldun[155]
A. M. Khazanov[156]
M. M. Kovalevsky
Phillipe Lucas and Jean-Claude Vatin[157]
Alan Macfarlane
Karl Marx
Margaret Mead[158]
Robert Montagne
Dr. Magali Morsy
H. Munson Jr.
C. Lévi-Strauss
G. E. Markov[159]
Germain Tillion[160]
Max Weber
Shelagh Weir

Philosophers/Theologians
St. Augustine
Joseph de Maistre
Hegel
David Hume
Luther
Frank E. Manuel
Nietzsche
St. Peter
Plato
Bertrand Russell
John Wesley[161]
A. N. Whitehead
Bernard Williams

Historians
Perry Anderson
Michael Cook
de Slane[162]
Edward Gibbon
Herder
Marshall G. S. Hodgson[163]
Robert Mantran
André Raymond

Paul Rezette
John Waterbury
Professor Montgomery Watt
Wittfogel[164]

Arabs and the Like
Agha Khan
Amir Abd el Kader[165]
Gaddafi
Grand Mufti of Central Asia
Hussein[166]
Ibn Khaldun
Khomeiny[167]
Ibn Weber[168]
Muhammad[169]
A Muslim Hegel[170]
Nasser
Osman dan Fodio
Qadi of Medina
Anouar Sadat
Shah (Pahlevi)
Omar Sharif
Yazid[171]

Historical Personages
Caesar
Charlemagne
Christ[172]
Pontius Pilate

None of the Above
General Daumas
Charles Henry Churchill[173]
T. S. Eliot
Friedrich Engels
Jehovah
C. G. Jung
Keynes
Thomas Kuhn
T. E. Lawrence
Napoleon[174]
Alexis de Tocqueville
Valentino
Oscar Wilde[175]

One thing is quite clear in this essay: Ernest Gellner is widely read and wants the reader to know it. No one could dispute this fact, but it is hard to see the relevance for the purpose at hand. There is nothing amiss in referencing other authorities, but surely a book on *Muslim Society* is oddly served by a Eurocentric intellectual and political who's who. Jehovah I can understand, not literally of course, but Bertrand Russell and Oscar Wilde make strange literary bedfellows indeed for explaining Islam.

In my mind the most compelling, and admirable, aspect of Gellnerian prose is the sheer delight in his spinning of phrases and bountiful supply of *bon mots*. Those who have read Gellner will hopefully appreciate the tribute in my own parodic representation; those who have not read Gellner for the fun of it are missing the British e[ru]dition of Clifford Geertz, perhaps a bit less dry and more Pythonesque. Consider the following gems:

> But the initial success of Islam was so rapid that it had no need to give anything unto Caesar . . . Hume was a Protestant Scot. He was also a man of the Enlightenment. There is no law against being either of these two estimable things . . . The motto of a proud soul, evidently, is not so much "No taxation without representation," but rather, "No taxation *at all*" . . . Sufism is then the opium of the people . . . The feud, like football, makes no sense if you do not know which team you are in . . . The Shi'ite deity is not so much a hidden god, as one given to playing hide and seek with men.[176]

There comes a point, one which I am constantly stretching, where being clever becomes a substitute for being correct, where rhetorical smoke has no fire in the belly of an argument. Certainly the point of no return is reached with Gellner's flippant disclaimer that the reader accept his model because any model is better than none.[177] The reader is ill-equipped to ignore the quips of the author, especially when they punctuate what might otherwise be a tedious and pedantic romp through overwrought philosophical prose. When Gellner makes us smile at how he writes, we are seduced into taking his satirical similes as theoretical insights. Islam, however, is only represented as a segment of the author's creative imagination.

Chapter 3

Beyond the Veil: At Play in the Bed of the Prophet

The entire Muslim social structure can be seen as an attack on, and defense against, the disruptive power of female sexuality.[1]

Fatima Mernissi

The Judaeo-Christian West has long orientalized the Muslim East through romanticized images of sensuous harem girls. More recently, a more secularized West still orients a more foreboding Islam through dimwitted, even Disneyized, media nostalgia of the *Arabian Nights*. From the black-and-white "I Dream of Jeannie" to the childlike coloring of "Aladdin," the "Oriental" woman is easily recognizable. In such ethnocentric views the Muslim woman is depicted as especially oppressed, screened off by the whim of tribalized males from her all too alienable rights as a modern female.[2] The various academic discourses on women and Islam have evolved from Orientalist to feminist on the outside and range from conservative apologetic to moderately confrontational on the inside. Western feminists have often defined Islam in the Middle East as a paradigmatic case of patriarchy, one of the more visible blips in the ongoing history of male domination. Yet many Muslims suggest with missionary zeal that Islam is the ultimate liberation for women. Not surprisingly, there is no dearth of recent literature, popular and academic, on gender and Islam.[3]

There are many ways to gender Islam, but all must eventually be linked to the prophet Muhammad as founder of the faith. The generally agreed-upon givens that the prophet of Islam practiced polygyny and that Islam is usually viewed as a sex-positive religion make the gender views attributed to Muhammad a central point in the ongoing debate over Islamic gender ideology. Without meaning to be facetious, it can be said that a major vantage point for gendering Islam has been in the bed of its prophet, even if it is more precisely a procrustean than a procreative bed. The issue here, at least from my standpoint as an anthropologist, is not historical reality, not even a form of literary criticism that would separate supposed fact from suspected fiction. The bed has been made and remade both by the followers of Muhammad, those who put together the traditions and wove the sacred history into a meaningful narrative at various points in the history of the religion, and by Islam's detractors. What Muhammad actually did and said is a moot point, unless you are a believer; what others make of all this is what matters, if you think like an anthropologist. Of primary interest, at least in the present climate of postmodernified anthropology, is how the views attributed to the prophet of Islam play a pivotal role in the continual discursive act of defining gender among as well as over and against non-Muslims.

For Muslims, the seventh-century prophet Muhammad serves not only as the divinely appointed messenger of the faith but also as the primary paradigm for practical, everyday behavior. Christian apologists, inflamed as much by political and economic concerns as theological purity, have tended to disparage Muhammad as a less-than-perfect prophet who distorted the earlier faiths of Judaism and Christianity to his own ends. The clincher in such holier-than-thoued argument was the alleged libido of Muhammad, whose views on sex, as nineteenth-century Orientalist William Muir put it, were "offensive to the European ear."[4] For a religious tradition like Christianity, which focused on asceticism as a defining characteristic of holiness, there was little acknowledged sympathy for a prophet who married more than once and accepted sex as an integral part of conjugal bliss. Given the intimate details of Muhammad's married life as recorded in Islamic texts, the grist of the Muslim's mill could easily be twisted through the ill-will of Western apologists. Muhammad "immoralized" readily became the apologist's trump card.

In order to situate the gendering of Islam in anthropological study, we need first to bring to the surface some of the prevailing gendered versions, including those in Western Orientalist writings. There is no need here to further document *ad nauseum* the historical tradition of cultural imperialism over Muslim populations. Accepting Edward Said's basic premise that "Orientalism" has more to do with "our" world than the Orient per se,[5] then how "we" gender Islam is as much a statement about how gender plays in our own minds as what Muslims think, say or do. Likewise, if we are willing to admit that nowadays "all cultures are involved in one another; none is single and pure, all are hybrid, heterogeneous, extraordinarily differentiated, and unmonolithic,"[6] then we should be all the more suspicious of any attempt to define a decisively "Muslim" concept of gender, whether or not the source material is medieval or modern.

An anthropological analysis of discourse is hardly novel. Such rhetoric has shaped much of the reflective writing in ethnography over the past two decades, so much so that the term "discourse" has only had to be intoned to validate theoretical closeness to the cutting edge. Whether as archaeologist of knowledge, in the Foucauldian mold, or as imaginer of communities, with a print-biased Andersonian angle, there is still a felt, disciplined need to look at the way we look at others, at the same time as we look at the way they look at others and at times even look back at us. This is not just a deeply played approach to interpreting over the shoulders of our informants as they read and act out their texts, but a more basic questioning of genre, style, and intent at a time when the meaning of content has been progressively devalued, de-emphasized and at times de-concocted.

My focus here is on how the Muslim prophet Muhammad is portrayed as the archetypal male in the creation of avowedly "Muslim" gender roles. I look first at the negative criticism of Orientalist scholars as exemplified in an influential biography, *The Life of Mahomet*, first published in 1861 by William Muir. This is tempered with a very different kind of romantic Orientalist discourse in Nabia Abbott's 1942 publication of *Aisha: The Beloved of Mohammed*.[7] For the positive valuation of Muhammad's view and treatment of the women in his life, I examine two twentieth-century Muslim viewpoints, the female Egyptian Quranic scholar known as Bint al-Shati' and the progressive Egyptian author Muhammad Haykel. Finally, I arrive at the discourse of Moroccan sociologist Fatima Mernissi, whose ethnographic data can be, and at times must be, filtered through the same debates and discursive turns that have engaged Orientalists and believers for several centuries. Also writing from a feminist perspective is Elaine Combs-Schilling in her *Sacred Performances*, which constructs a view of Muslim gender from ethnographic and historical research focused on the ongoing sultanate of Morocco.

This is not a survey of who said what and why. My foray is more pointedly a sharp counter-cut through the rhetoric on Muhammad as gender maker. The present study is not prolegomenon to reconstruction of a historical Muhammad or of a viable "Muslim" view of gender. I do not argue that these are unworthy goals, but my interest here is in the process by which ideas about the life of Muhammad enter into the rhetoric defining Muslim gender or, as Mernissi subtitles her earlier study, "male–female dynamics" among Muslims. Islam has generally been gendered on what its prophet is said to have said and done rather than what Muslims in various lived-in contexts actually do and say they do. While those who conduct ethnography among Muslims need not make any overt reference to Islamic texts and their interpretation, admittedly any such anthropologist would invariably have been exposed to and somewhat influenced by the kind of rhetoric I wish to explore here. I start with my own bias that, of all the ethnocentric baggage we carry into the field, our gender models are invariably the most overweight.

The rhetoric of gender discourse needs to be explored in a way that helps us better understand how our cumulative, and often competing, images of Muslim male and female are shaped. I have purposefully selected texts with discernible bias, con and pro. I situate these texts less in their ideological context, including the ethnocentric culture specifics beneath a homogenized global imperialism, than within the commonly shared *cul-de-sac* of a male-ordered gender bias. Each selected text packages Muhammad's views of women and sexuality, but the trope *du jour* is the rhetoric regarding a critical, gender-defining event in Islam: Muhammad's marriage to his adopted son's wife, Zaynab, and the subsequent "descent" of the "veil" passage in Quranic revelation. Muslims believe that Muhammad not only married his former daughter-in-law, but that God commanded him to do so. This is presented as hard evidence for the anti-Muslim apologist's prosecution that Muhammad was immoral and the Quran mere self-serving propaganda. Beyond this, the visible fact of an exotic veil in Muslim societies is without a doubt the preeminent symbol in the Western imagination of Muslim gender. The penchant for unveiling this female for a Western gaze is equally entrenched in those academic disciplines that concern themselves with Islam. Having played with the ideas the texts play with, I suggest some of the discursive links I see embedded in the respective discourses that gender Islam through these specific texts.

Orientalists

Mohammed was now near three score years of age; but weakness for the sex seemed to grow with age; and the attractions of his increasing harim instead of satisfying appear rather to have stimulated desire after new and varied charms.[8]

William Muir

Mohammed, the prayerful and perfumed prophet of Islam, was avowedly a great lover of the ladies, for whom, in turn, he held no small attraction.[9]

Nabia Abbott

Few scholars after Edward Said's seminal *Orientalism* would argue that academic discourse on the Orient is innocent. One of Said's candidates for a clearly biased Orientalist scholar of the nineteenth century is Britisher Sir William Muir, whose *The Life of Mahomet* was for half a century the "definitive" biographical portrait of Muhammad in English. This was first published in four volumes in 1861, then went through a second (1876), third (1894) and even revised fourth edition (1912) after his death. Muir's scholarly parsing of the life of Islam's founder draws heavily on the biographical details recorded by the early Muslim historian al-Tabari (died 923 C.E.) and a few other textual authorities, who are footnoted on virtually every page. Muir as a somewhat less than dispassionate Western historian frequently comments on the rationality of Islamic beliefs and claims about the prophet, as well as aspects of Muhammad's life of which he either approved or found wanting and, at various times, even debased. His stated goal, after lamenting the fictitious basis of much of the Islamic commentary, is that "by a comprehensive consideration of the subject, and careful discrimination of the several sources of error, we may reach at the least a fair approximation to the truth."[10] Of course, it is Muir as the self-professed objective non-Muslim who lays down the principles "separating the true from the false in Mohammadan tradition." Today the blatant bias of this text stands out as vividly as the archaic Victorian prose. What was once a major academic text on Muhammad has been mercifully relegated to rare-book-collection status. Yet, one edition or another is still available in many university libraries and occasionally consulted by naive students charged with writing term papers on Islam.[11]

Muir as scholar does not shy away from the issue of Muhammad's character, a characteristic concern of overtly Christian readings of the founder of Islam. We are introduced time and time again to a prophet who appears to be anything but a prophet, but rather someone "of a highly strung and nervous temperament."[12] Nowhere is the character of Muhammad more pronounced, in Muir's oblate discourse, than in his relation to women. This is hardly surprising, since Christians had longed used this one issue as the most damning in its rejection of Islam as apostasy. Early on in his biographical portrait, Muir informs us that it was the forty-year-old "comely widow" Khadija, Muhammad's employer in trade, who "cast a fond eye upon the thoughtful youth of five-and-twenty."[13] "Nor," we are told, "when he departed, could she dismiss him from her thoughts." Not unlike a formulaic nineteenth-century romance, Muir adds that "at last her love became irresistible, and she resolved in a discreet and cautious way to make known her passion to its object." Sending her sister, she lets Muhammad know of her desire for marriage and then manages to get her father into a Lot-like drunken stupor so he will not object to a marriage with someone who must have seemed at the time not to be a very desirable match. Of course, for an apologist from a religion whose founder arrives via virgin birth, the recorded facts about Muhammad's "connubial state" are suspect from the start. Further on, Muir returns to Muhammad's marital life to discuss wives number four and five and expand upon the issue of "the veil." Muir contends that although Muhammad "had been content with the three inmates of his harem" for over a year, he added two more wives in rapid succession due to his "weakness for the sex." The explanation that most of these marriages were political does not impress Muir, who attributes to the seventh-century prophet of Islam the libido, disguised or not, of a Victorian dandy.

Muir makes his own feelings explicit, to the extent this was not obvious already, in his treatment of Islamic marriage laws, which deeply offend his Christian values.[14] He was specifically upset by the open discussion of sexuality in the traditions, even wondering "how any translator can justify himself in rendering into English much that is contained in the Sections [of tradition collections] on marriage, purification, divorce, and female slavery." Muir is further aggravated by Muhammad's polygyny. While admitting that the number of wives is limited to four in Islam, he notes that divorce is easy and quick and that female slaves are common; thus, it is as though there is no "moral" concept of marriage at all in the religion. Female slavery is so endemic to Muslim communities, Muir avers, that it "will hardly ever be put down, without alien pressure." The female slave, Muir continues, is the "toy of her master, sported with at his pleasure, or cast unheeded aside." This is an ironic comment coming from a book first published in 1861, when such toying was still the law of the land in the southern exposure of Britain's former American colony. Need one wonder why Muir's rhetoric rated inclusion in Said's documentation of Orientalism as complicit in the moralizing service of empire?

As to the Muslim claim that their religion elevated and improved the status of women, Muir agrees and disagrees. Muir praises, grudgingly it would seem, the law of female inheritance and the right of a woman under Islam to refuse a marriage partner. Yet overall "the condition of woman is that of a dependent, destined for the service of her lord, liable to be cast off without the assignment of any reason and without the notice of a single hour."[15] To claim that Islam in any way liberates the female requires, in Muir's none-too-terse prose, that "we put aside the Veil and the depressing influence which the constraint and thraldom of the married state has exercised upon *the sex at large*."

The image of a subjugated female slave as the archetype of the Muslim woman resonates well with the casual bias of then contemporary male travelers to the Orient.[16]

Before leaving the subject of Muhammad and his wives, Muir mentions the Quranic verse that gives Muhammad the right to visit Aisha, his favorite wife, as he wished rather than in the fair sharing of equal sexual treatment the prophet had suggested for his followers.[17] The patronizing tone of Muir is particularly evident at this point, as he complains that such a trifling command should be recited, as part of the continuous course of Quranic recitation, in every mosque throughout Islam. He sums up the character issue with the concluding remark that "We gladly turn to other matters" as he turns to a political raid of Muhammad. Yet Muir, gladly or otherwise, returns to the subject of Muhammad's marital problems only four pages later in the same chapter. This is the issue of the "misadventure" of Aisha, Muhammad's youngest wife. In this case Aisha is accidentally left behind in the prophet's caravan and brought back to camp later by a man named Safwan. Muir comments with obvious disdain that "the scandal-loving Arabs were not slow in drawing sinister conclusions from the inopportune affair, and spreading them abroad."[18] This, Muir informs us, is the reason that the Quran (surah 24) demands 24 stripes, the penalty for fornication, for those who falsely accuse a married woman of adultery. Once again, Muir is at pains to show how the Quran was manipulated at will by Muhammad to suit his base domestic instincts.

Where Muir demurred at the romance of Muhammad, Nabia Abbott reveled in "a perfumed prophet" who was "avowedly a great lover of the ladies."[19] No less a scholar of stature than her staid predecessor, Abbott wrote not as an overt detractor of Islam but as one sympathetic to "progressive Moslems of today, be they Arab or Persian, Indian or Chinese, Mongol or Turk, [who] not only are keenly interested in the problems of the current Moslem woman's movement but show a gratifying curiosity regarding the achievement of the historic women of Islam."[20] This is not to say that Abbott was less than dispassionate, but she saw the progressive intellectual change in the Islamic world of her time as a positive development. This was also a time when the direct colonial control of Islamic countries was on the wane and Islamic "fundamentalism" had yet to be defined or styled as a threat.

Abbott, as a scholar not at obvious odds with the tradition she studied, was drawn to the story of Aisha, the daughter of the prophet's close confidant Abu Bakr. After the death of Khadija, Muhammad's first and only wife for twenty-six years, the nine- or ten-year-old Aisha "with her lively temperament and pert charm, brought a refreshing air of romance into the closing years of his life."[21] Abbott did not approach Aisha as the immature lover of a dirty old man; Muhammad was better regarded as an "elderly" man who understood Aisha was still a child playing with toys and thus he "let nature take its course."[22] In retelling the sacred history of the prophet's marital life, Abbott was quick to observe how the same event would engender criticism from extremist anti-Muslims and at the same time reverence from devout believers. She surmised, perhaps ahead of her time, that at the time of Muhammad sex "was nearly an obsession with the entire population, and sex talk, frank among the better element, tended to be indecent and lewd among the worst sort."[23] In the context of what she perceived as pre-Islamic sexual mores, institutions such as the veil and harem in Muhammad's own life appear as reformist rather than reprobate. "It is easy enough to overemphasize the jealousies and discords inherent in the harem system, even to the point of leaving the impression that harem life, for the women at least, is one bitter and continuous competitive struggle devoid of any peace or friendship," Abbott cautions.[24] And, it "is equally easy to paint the master of the harem as the sensualist utterly lacking in finer family sentiment." Easy but wrong, since she recognized "an unhappy tendency among some Western biographers of Muhammad" to blame the prophet for "his pronounced and avowed weakness for the fair sex."[25] For those who blame Muhammad for an extraordinary sex drive and penchant for multiple wives, she reminds the reader of Joseph Smith and Brigham Young and the respectability finally achieved by the Mormons in the American West. Unlike Muir, she did not see the monogamous ideal in the West as intrinsically more moral in practice than Islamic provisions for polygyny.[26]

This brief introduction to the work of two scholars trained as Orientalists and specializing on the life of Muhammad illustrates lines of cleavage that are often obscured in the broader deconstruction of Orientalism as such. It may have been useful a few decades ago to speak in terms of a "worldwide hegemony of Orientalism" in order to help correct past and ongoing injustices, but cross-cutting an alleged "ism" as broad and politically charged as the intellectual universe of those who study the geographically indistinct Orient are real differences in intentions, prejudices, and rhetorical skills.[27] To be sure Western scholarship has aided and abetted the colonial enterprise on the whole, but this is hardly sufficient cause to ignore the vast academic service provided in historical, literary, and grammatical analyses, or the utilitarian value of translations of basic "Oriental" texts. Orientalist Muir spoke to an audience that at the time shared his ethnocentric bias against Islam, but Orientalist Abbott wrote against the prevalent stereotyping entrenched in her own intellectual heritage. Perhaps the postmodern de-mything of Orientalism is more mythical than is admited in the heat of politically correct polemics.

Believers

If women owe any gratitude to any one on the earth, it is to Muhammad.[28]

A. Rahman

May God forgive those who claim that his Prophet's heart did not throb with the love of Aisha, nor that it was attracted to Zaynab bint Jahsh! May God forgive those who claim that his feelings had no part in his marriages.[29]

Bint al-Shati'

Muslims are taught to respect every recorded word of their prophet. These traditions of what Muhammad said, along with amply recorded aspects of his behavior, provide the grist which generations of believers have mulled over in a wide variety of cultures, including our own. While no visual image of Muhammad is permitted in Islam, Muslims in a sense strive to create themselves in the verbal image of their prophet. He is, in a word or two, the complete man (*insan kamil*).[30] While Muhammad may be viewed doctrinally only as a mortal man, he is as good as any man could hope to be; and, by extension, as good as any woman could hope to marry. Muhammad, prophet and man, rests on a pedestal among Muslims as the ideal husband. Beyond this, it is argued that he raised the status of women at a time when women were otherwise treated, especially in the barbarian-dominated West, as little more than property or chattel. Typical among apologists for the faith is the sentiment that a Muslim woman should be thankful to God for her liberation through Islam.

To speak of traditional Islamic views on gender, it is important to realize that "gender" as such was not an independent issue in the way it has been shaped in Western perspectives since the influence of twentieth-century feminism. When earlier Christian apologists descried Islamic sexual morality and vilified the prophet, this was hardly a result of any enlightened views on gender equality in their own cultural tradition. Nineteenth-century scholars such as William Muir were as embedded in gender bias as they asserted Muslims to be. While it is true that Islam, following Judaism and Christianity, does not recognize any significant female prophets, there are significant women in the sacred history of the religion. Most notably, the Quran addresses matters related to the prophet's wives, who are honored with the title "mothers of the believers" in numerous places.[31] Consider the legacy of Aisha, Muhammad's favorite wife, who is credited with relating many of the "sound" traditions of the prophet, at least in Sunni perspective. Thus, Muslim commentators over the vast stretch of the religion's history have not seen the gender issue as a relevant criticism of their religion. Not surprisingly, Muslims tend to react to negative Western rhetoric, apologetic or feminist, as misinformed and malicious.

One of the more important "modern" interpretations of the life of the prophet Muhammad is Muhammad Haykel's 1935 reworking of the biographical material for the modern age. For the Egyptian scholar Haykel, steeped in a milieu of blatantly biased attacks by Western scholars, this was a time when "Christian fanaticism against Islam continues to rage with such power in an age which is claimed to be the age of light and science, of tolerance and *largeur de coeur*."[32] Haykel was not alone among Muslim opponents of Western depiction of their religion, but it would be another six decades before an "Oriental" scholar, namely Edward Said, would lay bare the dimensions of this fanaticism in a critique of Orientalist discourse that would really catch on. Haykel attributes the hostility of Christianity against Islam to Western ignorance, especially regarding the life of Muhammad. He offers a basically orthodox view of the sacred history of Muhammad, although it is his stated aim that "this will be a scientific study, developed on the western modern method, and written for the sake of truth alone."[33] This is not the work of a mindless conservative, ignorant of the achievements of Western tradition. However, this is very much a polemical defense of an Islam under siege by Christianity. The bottom line for Haykel is that since Islam as such is so "sublimely noble, simple, and easy to understand," its critics are reduced to the trick of shifting attention from the idea to the person advocating it. Hence, Orientalist attacks on Muhammad result from the alleged fallacy of an *argumentum ad hominem*.[34]

Haykel takes to task those Orientalists who summarily reject the Quran as revelation. Noting that only a few of these scholars have branded the Quran as a deliberate forgery, Haykel turns to none other than Sir William Muir, whom he recognizes as "a missionary who never misses occasion to criticize the prophet of Islam or its scripture."[35] Indeed, some ten pages of Muir's text are quoted in large part to prove Haykel's point that seemingly respectable Orientalists cannot really deny the authority of Islam's defining text. This argument is presented in the preface to the second edition as a response to criticism from an Egyptian Muslim whom Haykel found "too ready to accept what the Orientalists say and regard it as true knowledge."[36] In an engaging rhetorical twist Haykel thus manages to quote the Christian devil to do Allah's work. It is debatable whether such rhetoric would have satisfied the unidentified critic, but it does establish Haykel's point that he is as qualified to present a "scientific" study of Muhammad as any Western historian.

Haykel's text does not dwell on Muhammad's view of gender apart from refutation of Orientalist vindictive on the prophet's morality. The match between forty-year-old Khadija and twenty-five-year-old Muhammad is glossed by Haykel as a logical outcome of Muhammad's business success and good character. The love of Muhammad for Khadija, Haykel avers, is "not the raging passion of youth which is as quickly kindled as cooled or put off."[37] It was thus the prophet's loyalty, truthfulness, and respect that led to this first marriage, not passion or lust. This explains the fact that Khadija was the first convert in Islam and steadfastly remained one of the prophet's main confidants. In two later chapters Haykel provides details on the wives of the prophet. The main thrust of his argument is that Orientalists have vented unjustly against Muhammad and that their outrageous claims can be refuted by logic. If, argues Haykel, Muhammad was faithful to his first wife Khadija for the twenty-eight years of their marriage, why would he at the advanced age of his early fifties suddenly start marrying out of unbridled sexual desire?[38] And if he were so attracted by feminine wiles, why was his second wife the widow Sawda, who was neither beautiful nor wealthy?

In his explicitly modernist—though not yet decolonized—guise, Haykel goes so far as to argue that Muhammad "stood for monogamy and counseled its observance."[39] This is a daring admission, but one which could effectively counter the Western claim that Islam denied equality to the female by its acceptance of multiple wives and the so-called harem complex. Not all Muslim intellectuals have been willing to concede this point. Among them is a female religious scholar named Aisha 'Abd al-Rahman, whose theological pen name was Bint al-Shati' (literally, "Daughter of the Beach"). This Egyptian scholar memorized the Quran as a child and in 1950 received a Ph.D. from Cairo University. She wrote a commentary on the Quran, and later published a book on the prophet's wives as a direct response to the rationalizing effort of Muhammad Haykel.

Like Haykel, Bint al-Shati' addresses the prejudices of Orientalist scholars in their attacks on Muhammad and the practice of polygyny. In her defense of Muhammad as the ideal man, she makes a virtue out of what others have seen as a vice. In reference to Muhammad's married life, Bint al-Shati' argues that "we are dazzled by its vitality which knows no sterile passion or fossilized affection. All of this because he behaved naturally and allowed his wives to fill his private life with warmth and excitement . . ."[40] To Bint al-Shati' there is nothing shameful in Muhammad's passion, nor should anyone deny his physical attraction to Aisha or the beautiful Zaynab. She even notes a tradition in which the prophet permitted any man to examine fully the woman he proposed to marry.

Of major interest in her account is a stirring defense of polygyny as a possible boon to a woman. First, she ridicules the sanctimonious monogamy mongering of Islam's critics, since adultery and loose morals are so widespread in the West. Then she disputes the idea that polygyny enslaves the Arab woman and reduces her to an object for man's entertainment. "In reality," asserts Bint al-Shati' "polygamy quite often placed upon a man a heavy burden and saved the Arab woman from a more degrading system—namely the modern slavery which recognizes only one wife and leaves other women to be carnally used and left without a place in society."[41] In a striking retort to reformists such as Haykel, she concludes that a "woman may contentedly prefer to have half of one man's life rather than the whole of another." It was understandable to her that several women would desire to have the prophet as their husband. Bint al-Shati' argues as well that the reported rivalry between certain of the wives was at base a petty possessive desire to have the prophet each to herself. He was, after all, "the" prophet.

Feminists

The Prophet said, "After my disappearance there will be no greater source of chaos and disorder for my nation than women."[42]

Muhammad, quoted by Fatima Mernissi

Islam did not invent patriarchy and patrilineality, but it did make them sacred.[43]

Elaine Combs-Schilling

It is difficult as an anthropologist to conduct ethnographic fieldwork in much of the world, no longer just the Middle East, without running into Muslims and their grievances. The ethnographic data for Muslim women were rather scanty up until the decades of the 1970s, when female anthropologists started publishing their accounts. The first major "anthropological" discussion of gender and Islam came with Moroccan sociologist Fatima Mernissi's *Beyond the*

Veil, originally published in 1975. Mernissi's text literally filled a vacuum and quickly was filed on most university and public library shelves. This explains in part its longevity into a revised edition in 1987. At the time of its first appearance there was no major ethnography, apart from museum-style descriptions, on gender in an Islamic context. Virtually all the previous information on the role of Muslim women, especially for the Middle East, came from travelogues stemming back well over a century. As Judy Mabro has documented in her selection of "veiled half-truths" from the travel literature, this tourist and pilgrim information was little more than quaint and more often blatantly ethnocentric.[44] Indeed, there were only a few perfunctory observations on gender by male anthropologists who had worked the region.

Most of the anthropological studies of gender and Arab or Muslim women focus, understandably, on contemporary cultural contexts. Mernissi dutifully presented her sociological data about modern Moroccan women, but her text became popular due to the extended discussion of "the traditional Muslim view of women" by a Western-trained feminist looking at Islamic texts. More recently, in *The Veil and the Male Elite*, she returns to the same theme to expand on her own sense of what these seminal religious texts communicate about Muhammad's view of gender. In another ethnographic study on Morocco, Elaine Combs-Schilling likewise extends beyond the contemporary data to an analysis of Muhammad and his wives as an integral part of her overall argument. Each of these studies appeared before historian Leila Ahmed's well informed survey of women and gender in Islam, Malti-Douglas's work on gender and discourse in Arabic literature and Barbara Stowasser's masterful documentation of Islamic source material on Muhammad's wives. For the earlier texts of Mernissi and Combs-Schilling, this is indeed unfortunate.

Until 1990 *Beyond the Veil* was by any measure the single most influential discourse on Islam and gender by a social scientist written explicitly for a Western audience. Translations of *Beyond the Veil* have been made into Arabic (1987), French (1983), Dutch (1985), German (1987), Spanish (1975), and even Urdu (1987).[45] Richard Martin and Mark Woodward are among those who elevate Mernissi as "a Muslim intellectual with a Western education, able to analyze and criticize Western thought on its own terms."[46] Yet, as Katherine Bullock observes, the farther scholars are from the actual study of Islam, "the more they take her word to be the 'truth' about Islam."[47] In its second edition, Mernissi remarks that her book "does not seem to age, because it is not so much about facts as data as it is about an ageless problem."[48] True enough, the ethnographic data she presents

for Morocco are sparse and rarely cited; it is her discourse on how
Islam defines gender that accounts for most readers' interest. This is
prefigured from the start in a book subtitled "Male–Female Dynamics
in Modern Muslim Society." Her original intent, as noted in the
introduction to the first edition, was to "explore the male–female
relation as a component of the Muslim system" and she speaks
throughout in a loose pseudo-Weberian style of "Muslim society" and
"Muslim women."

The first part of the book treats "The Traditional Muslim View of
Women and Their Place in the Social Order." Following the title
quite literally, Mernissi proceeds to lift the veil and allow the reader
to gaze at "the" Muslim everywoman. This clearly blends well with a
background of Western voyeuristic interest in the mysterious Orient,
especially when the first two paragraphs of the first chapter compare
an eleventh-century Islamic scholar's view on sexual instincts to
Freud's concept of the libido.[49] We are further treated to this almost
millenium-old Islamic authority, al-Ghazali, for his explanation of
why God created man with a penis and testicles and how sexual desire
is "a foretaste of the delights secured for men in Paradise." All this
sets up a notion that constitutes the ultimate f-word for Mernissi, the
idea of *fitna* (literally, "disorder" or "chaos"), which she further
glosses as "a beautiful woman," the connotation of a *femme fatale*
who makes men lose their self-control. Dialoguing Western social
scientist Freud and Islamic theologian al-Ghazali, Mernissi ultimately
drives to her main point, that the "entire Muslim social structure can
be seen as an attack on, and a defense against, the disruptive power of
female sexuality."[50] Electra in Marrakech.

"Why does Islam fear *fitna*? Why does Islam fear the power of
female sexual attraction over men?" asks Mernissi?[51] The essentialization
of female nature by the Arabic term *fitna* is a pervasive trope in her
text. A reader might assume that this is a primary meaning, despite
Mernissi's gloss of the term as "disorder or chaos." In fact *fitna*
occurs in the Quran over thirty times with a range of meaning covering temptation, trial, persecution, apostasy, and treachery.[52] The root
sense of the word deals with burning or smelting with fire, hence the
translatable connotation of trial by fire. Nowhere in the Quran is
woman referred to as *fitna*, although wealth and children are.[53] In
the early formation of the Islamic community the term was used for
the civil strife and battles surrounding the battle for power between
'Ali and Mu'awiya. It has been used for all kinds of conflict, spiritual
as well as political, which afflict Muslims. The Arabic lexicons rarely
mention its figurative application to women, although such usage
appears to be common in Moroccan dialect. The fit of *fitna* with

female is sealed with Mernissi's citation of a tradition in which Muhammad allegedly said women would be the greatest source for chaos and disorder among Muslims.[54] Surely, given the political trials throughout the history of Muslim societies, it is absurd to take such a claim literally. Egyptian anthropologist Fadwa El Guindi is right to accuse Mernissi of "reducing a complex sociopolitical structure purely to gender and sexuality."[55]

Having argued that Islam is at least anti-"female sexuality," if not anti-female in a broader sense as well, Mernissi suggests a "fundamental discrepancy" in "Muslim sexuality as civilized sexuality."[56] The regulating or constraining of female sexuality is said to be inherently unfair since the institution of polygyny maintains promiscuity in male sexuality. It is at this point that Mernissi explores the implications of the life of Muhammad and argues forcibly that "the virtually hysterical attitude of Arab-Muslim leaders to the emergence of female self-determination" is due to the "Muslim time-frame" of social conditions at the time of the prophet in the seventh century. The assumption here is that nothing is new under the Middle Eastern sun: "fourteen centuries seem to have elapsed without major upheavals or fatal discontinuity, and the future promises to be a continuation of the past."[57] Like the Orientalists and bibliophiles of a previous century, the idyll of a Middle East with little or no cultural change resonates in Mernissi's writing. From here on the discussion turns to her research among urban Moroccan women in the early 1970s, followed by a short conclusion on "Women's Liberation in Muslim Countries." To note that the data part of the book is anticlimactic is an understatement.

Having read this book when it first appeared, while I was a graduate student preparing to conduct ethnographic research in the Middle East, and then more recently after the second edition, I am prone to agree with Mernissi that the text does not seem to age. Fortunately, I have. This is in no small part due to my own ethnographic research among Yemeni villagers for over a year, but also to a continuing interest in the Arabic texts recounting Islam's sacred history. I suggest that the assumptions upon which *Beyond the Veil* is constructed are as out-of-date today as they were seductive when they first appeared in the mid-1970s. At issue is not that this book was written as a feminist challenge to Islam, but that it ultimately rigidifies biased stereotypes about Islam and Arabs, endemic stereotypes that have more recently been effectively countered in other feminist scholars' work. Several reviewers of the original edition were aware of the polemical nature of Mernissi's argument and her shaky assumptions of a universal and unchanging Islam.[58] But, by and large, these "shortcomings" have been glossed over because, as Eickelman notes, Mernissi "poses some interesting questions."[59] But do the questions really matter when the book presents simple-minded ahistorical answers that readily appeal to an ethnocentrically liberal Western readership?

My primary problem with Mernissi's lifting of the veil is that the Muslim woman we are invited to gaze at is a disembodied ideal, more properly here an un-ideal. The text as written does not confine itself to Moroccan views of Islam or female sexuality but purports to speak of "Muslim Society," the "Muslim East," the "Muslim system," the "Muslim order," and even "Islam" as such. Throughout the first part of her book we do not see one actual Muslim female, only an artificial image built up of select passages from a single Islamic scholar and various statements attributed to or about Muhammad. We are essentially presented with an indictment that says here is the proof that Islam is a sexist religion, something about which the average Western reader needs little convincing. The book's stated intent is to contrast "the way women are treated in the Muslim East with the way they are treated in the Christian West,"[60] but in fact this is attempted, as Mernissi freely acknowledges, without filling in the data. Nor is there even the pretense of an attempt to discuss gender in the "Christian West" apart from an uncritical acceptance of Freud. Mernissi, as social scientist with life experience in Morocco, contextualizes the limited field data she does include within a framework that is trumpeted as valid for all Muslims. I would not presume to comment on her conclusions about Morocco, where I have not worked, but I do find it odd that her statements about a constructed "Muslim" view of sexuality are commonly billed as a "classic study" when there is no evidence they extend beyond the views of a Western-educated, urban Moroccan who interviewed a number of women in the summer of 1971.[61]

Mernissi constructs a single dominant view of sexuality among Muslims while she purports to be doing sociology or anthropology. In one chapter Mernissi gives us the "Muslim concept of active female sexuality" and in another "the regulation of female sexuality in the Muslim social order." Would it make anthropological sense to speak of a specifically "Christian" sexuality, or something as absurd as a Jewish "social order"? Do Christians or Jews act as mere clones of a pervasive and universally valid view of sexuality, no matter what the social context? Or is it Mernissi's contention that only Islam is monolithed in stone by an overarching patriarchy? We are urged to think that in Muslim society there is a "structural dissymmetry that runs all through and conditions the entire fabric of social and individual life . . ."[62] All Muslim societies? Why not in all societies?

Ruth Benedict could hardly have patched together such an ideal pattern better, but she did so a number of decades earlier, and at least Benedict recognized a role for shreds of culture.

The reader is essentially being told that there is a determinative view of female sexuality reinforced on Muslims everywhere by a male elite. This is the trope Mernissi consistently applies in all her writing and for which her first book is most appreciated. In the entire first part of *Beyond the Veil*, the most glaring omission is of any specific cultural context; nor is there any attempt to provide comparative data from other Islamic societies. Islam is reified in a take-it-or-leave-it sense that fits squarely with the age-old preconceptions Westerners have about Islam and grates to the core a large majority of Muslims, female as well as male. The numerous ethnographies available since the publication of the first edition argue so convincingly against such an unfounded universalist view of gender among Muslims that it hardly seems necessary to continue to state what most anthropologists now take as obvious.[63]

In constructing her view of how Islam defines gender, Mernissi has frequent recourse to the prophet Muhammad, especially in her later texts, *Le harem politique* and *The Veil and the Male Elite*. In *Beyond the Veil* extensive coverage of Muhammad's marital life follows on her discussion of polygamy and repudiation. After noting that Muhammad was married to Khadija in a "monogamous marriage" for twenty-five years, Mernissi takes up the issue of several women reputed to have offered themselves to the prophet. We are advised that the custom of a woman offering herself, without family intercessors, was a pre-Islamic custom outlawed after the death of Muhammad. "If he [Muhammad] was the last Arab man to be chosen freely by women," Mernissi remarks, "he was also probably the last to be repudiated by them."[64] At this point in her argument several recorded cases are mentioned in which women broke off an arranged marriage to Muhammad. We are told that the general explanation of this among Muslim scholars is that they were tricked by jealous co-wives. Mernissi rejects this as "the work of Muslim historians who thought it necessary to disguise the embarrassing fact that the Prophet had been rejected and 'repudiated'."[65] For this twentieth-century author it seems more likely that each of these women would not want to marry an old man in his early sixties or share him with at least nine co-wives. The point that he was a charismatic political leader apparently did not counter a certain lack of sex appeal. If Mernissi is correct,these women were looking to bed Don Juan rather than Solomon.

Having rhetorically branded the generic male Muslim image through the foundational role of Muhammad as a dirty old man with roving eyes for young and attractive wives, we are then told that "the explanation of their [the women who rejected Muhammad] behavior is secondary here."[66] But in fact this is the primary focus and the novel information in her account. The average Western reader would not question that a young and attractive woman would be turned off by an old man in his sixties. In our commercialized and youth-oriented society this is not a romanticized love match. Of course, it happens all the time for the right, largely economic, reasons. But is it fair scholarship to read modern romantic ideals, even if they can be construed as Moroccan, onto a sacred history of events alleged to have occurred over fourteen centuries ago? Echoing Haykel, if age were such a critical factor in marriage choice at the time, what about a young bride like Aisha, whom Mernissi praises for her sincere love and devotion to Muhammad? And, to what extent does "love" as defined in modern cinema enter into arranged marriages with obvious political ramifications? The issue is not that Muhammad would not have been a romantic match, a point vigorously made by Bint al-Shati', but that Mernissi fails to ground her portrait in what the Islamic sources relate about Muhammad's loving behavior with his wives.

Another Moroccanist, Elaine Combs-Schilling takes up Mernissi's rhetorical flame in addressing the link between Islam, sexuality, and sacrifice. "As Mernissi (1987a) forcefully argues," Combs-Schilling reminds us, "Islamic sexual culture emphasizes the female as a powerful, seductive temptress who—consciously or unconsciously—is driven to capture the hearts and souls of men and bind them to her, interfering with the male's ability to focus on God."[67] Quoting Mernissi at the same pace others might be tempted to cite Foucault, the Islamic female is once again reduced to the *femme fatale*, a view we are told is "well-developed in the literature" or *fitna*, which is narrowly defined by Combs-Schilling as "the Arabic word for a beautiful woman."[68] Like Mernissi, Combs-Schilling dwells on the exotica that has long appealed to the uninformed Western imagination: heavenly sirens, which give the male in paradise a twenty-four-year orgasm,[69] a "dominant sexual culture of Islam" that "systematically works against the heterosexual bond,"[70] a tradition that says three of the best things on earth are perfume, women, and ritual prayer,[71] and rhetoric *in flagrante delicto* on the *Arabian Nights* scenario of a sex-crazed caliph who wedded, bedded, and beheaded one flirtatious *fitna* after another. The bottom line here is the author's essentialization of Islam in the two concepts of "patrilineality and patriarchy."[72] Thus, females are little more than "dead ends for their patrilines,"[73] somewhat alarming news for all the sayyids who claim descent through the prophet's daughter, Fatima. The Quran is said to legitimize male dominance as it is on earth and shall be in heaven.

Such an idealization of gender under the rubric of a normative patriarchal Islam is necessary in order for Combs-Schilling to develop her psychoanalytic interpretation of Muslim sacrifice. Here she revisits a paradigmatic sacrificial myth shared by Judaism, Christianity, and Islam, namely God's call for Abraham to sacrifice his heir, Isaac for some and Ishmael for others. Her characterization of this sacrifice is worth quoting in full:

> Yet Islam then makes the cosmic intercourse an all-male event: the erect father standing on the mountaintop in response to the command of a male-imaged God thrusting the knife towards the reclining son's throat in order for divine connection to be made, and then actually plunging the knife into the ram's throat. The plunging of the knife into the animal's white neck bears stark similarity to the plunging of phallus into vagina. And it serves an analogous but higher purpose—birth into transcendence, as opposed to birth into this world.[74]

In what I take to be a penetrating Freudian sense, the ritual act of slaughtering a sheep with a knife is thus reduced to the wet dream of an erect male thrusting his divinely perspired member into a subservient and victimizable vagina substitute. As hard rhetoric, these few lines manage to connote what Orientalists were usually too shy to denote, except occasionally in Latin. As I read this pruriency-prone passage, God and Abraham merge into a homosexual pair and the stand-in for the female is either a young child or an innocent but dumb animal. This ultimate cosmic rape scene thus serves symbolically as a fitting trope for male dominance through patriarchy. The gist of this just-so analysis, communicated with no apparent fear and trembling, not only genders Islam in a single, well-aimed blow but friezes the patriarchal image as a pervasive structural motif whenever Muslims are to be thought about.

A fellow ethnographer of Morocco, Laurence Rosen, finds the theoretical frame of Combs-Schilling "not only internally inconsistent and unfalsifiable but dependent on several metaphors that she has taken literally."[75] John Bowen contrasts the Moroccan example with research in Sumatra, arguing that Combs-Schilling goes too far in making the Moroccan sacrificial rite a synecdoche for all Islam.[76] One of the few admirers of this sacrificial plot is Carol Delaney, who defends Combs-Schilling's blunt attack on patriarchy as "a rich and convincing account of the meaning of blood sacrifice in Moroccan Islam and its relation to the story of Abraham."[77] It is unclear how many Moroccans, especially those who take the story literally as part of their faith, would be convinced. Since the symbolic connection is made by the anthropologist with no clear data from ethnographic dialogue or Islamic texts, I question what makes such a charged reading "rich." I am more prone to see the claim that the act is "sexual" stemming from the influence of an earlier and poorly argued work by Abdelwahab Boudhiba, who sensationalizes the Muslim act of procreation "in the form of an immanent thrust in which God himself participates."[78] Krishna in Marrekech.

Both Combs-Schilling's *Sacred Performances* and Mernissi's *Beyond the Veil* attempt to contextualize contemporary ethnographic data by first establishing a reductionist view of Islam and gender. Unfortunately, the generic model that each explicates is so pervasive that it overrides the cultural diversity of individual Muslim behavior observable across space and thinkable over time. Neither author provides ethnographic support from other Muslim societies; the Moroccan microcosm suffices for all Muslims. Nor do they critically acknowledge the diversity of opinion within the historical tradition branded as Islam. While Mernissi is able to sift through original Arabic texts, Combs-Schilling relies exclusively on secondary sources and translations, or so her limited bibliography would indicate. For the latter, it is somewhat ironic that an avowedly feminist work would quote Orientalist scholars such as Carl Brockelmann, whose name is misspelled in the text,[79] and flawed, nonscholarly treatments of gender such as the mondo-Freudian *Sexuality in Islam* romp of Abdelwahab Boudhiba.[80] At one low point Combs-Schilling compares the idea that "Muhammad took from nine to thirteen wives" to the numerical similarity that "Jesus took twelve disciples."[81] Unintentional as overt apologetic, this kind of statement still serves to reinforce the medievally Orientalist dichotomy of an Islamic preoccupation with sex in contrast to a Judeo-Christian/Western concern for religious devotion.

The Scandal and the Veil

And if you ask them [the women] for a thing, then ask them from behind a hijab. That is purer for your hearts and their hearts. And it is not for you to cause annoyance to God's Messenger, nor that you should marry his wives after him. Truly this with God would be enormous.

Surah 33:53

Veiling is a variable, not a constant, and no single fact about persons accounts for this variation.[82]

Jon W. Anderson

The most compelling symbol of a so-called Muslim gender model is certainly the veil, *hijab* in Arabic.[83] While it is clear that Muslims did not invent the veil, this visible distinction between male and female in traditional Middle Eastern societies is the focal point for most Western discussion of gender in Islam. The occasion for the Quranic verses that impose the "veil" on the wives of the prophet, and by extension to other Muslim women, is generally tied to Muhammad's marriage to Zaynab bint Jahsh. Zaynab was Muhammad's first cousin, a preferred marriage partner in this sense, but she was already married to Muhammad's adopted son, Zayd. The reason for Muhammad's delayed sexual interest in his cousin is usually linked to his seeing her naked or nearly so in her tent. Several verses in the Quran (surah 33:37–40) directly address this unusual case and legitimize Muhammad marrying his adopted son's wife as something God himself had decreed.

Whatever one may think of Muhammad's reputedly unreputable views on gender, this story is in many respects the most critical. My interest is not in the historicity or credibility of the story, but rather how it plays to both the detractors and defenders of Islam. As Haykel laments, in regarding this marriage "Orientalists offer their highest condemnation, in chorus with the Christian missionaries."[84] This was one of the most popular stories for Christian diatribes on the hypocrisy of Muhammad, who was villified for justifying his adultery and lust through a pretended revelation.[85] Thus, this controversial story is an appropriate example for comparing the rhetoric of the various authors cited above, especially in light of its relevance to the origin of the veil.

Orientalists did indeed relish this scandal in the life of Muhammad. As Muir reconstructs the scene, Muhammad went to visit Zayd, who was absent, and saw his daughter-in-law "in her loose and scanty dress" and "the beauties of her figure through the half-opened door had already unveiled themselves too freely before the admiring gaze of Muhammed." Zaynab recognized the prophet's fascination with her beauty and told her husband, who went to his adopted father, offering to divorce his wife. Muhammed's counsel to Zayd not to divorce Zaynab, according to Muir, came from "unwilling lips." As Muir relates, Muhammad at first hesitated in marrying Zaynab, lest there be a scandal, for even in pagan Arabia such a union with one's adopted son's wife was unlawful. "The flame, however," explains Muir, "would not be stifled; and so, casting his scruples to the winds, he resolved at last to have her." As though this was not scandalous enough, Muhammad "had to fall back upon the Oracle" to justify his lust as God's will.

At this point Muir turns to the veil as a means of seclusion enjoined upon Muhammad's wives for what to him were obvious reasons, given the preceding account of primal Arab lust.[86] After all, Muir argues, Muhammad's followers "could hardly expect to be freer from temptation than the Prophet himself." It is from the practical restrictions placed on Muhammad's wives, Muir continues, that similar "stringent usages" of the harem and seclusion evolved in Islam. So "the Mothers of the Faithful" were condemned to the harem in the Quran due to the original sin of the founder of the faith. "However degrading and austere these usages may appear, yet with the loose code of polygamy and divorce some restraints of the kind are almost indispensable in Islam, if only for the maintenance of decency and social order." This commentary is highlighted in the fourth edition by the marginal heading "Restrictions rendered necessary by loose code of Kor'an." In a footnote, Muir reminds his readers that "European manners and customs in this respect would be altogether unsuited to Mohammedan society."[87]

Nabia Abbott is well aware that Western spins on the Zaynab story give offense to Muslim scholars. In relating the "essential facts of the story" Abbott situates Zaynab merely "in light disarray" when Muhammad accidentally saw her in the tent. Yet the prophet did not ogle her but hurriedly "went away murmuring 'Praised be Allah who transforms the hearts!'" After this event, humble and unattractive Zayd experienced "no peaceful living with the haughty and ambitious Zaynab," whom he divorced despite the prophet's counsel not to do so. Unlike Muir, however, Abbott offers no condemnation of Muhammad, but rather criticizes the overly idealistic motives of both detractors and defenders. For Abbott, Muhammad is neither a hypocritical "voluptuary" nor a pure "saint."[88] She agrees with the moderately favorable assessment of the German scholar Tor Andrae that who at the time would question God giving Muhammad a few extra privileges denied to the common man.[89] Later, however, she argues that it was as much a burden as a blessing for Muhammad to have so many wives, since their constant bickering was a continual thorn in his side. Domestic life became such a squabble that at one point Muhammad "exasperated, retired from all his wives."[90] If we are tempted to search for a Western literary metaphor, this sounds more like Walter Mitty than James Bond.

Where Muir saw lax morals that pervaded both pre-Islamic society and the emerging community of believers, Abbott applies an exegesis to the veil passage that separates the introduction of the *hijab*, which she translates here as "curtain" rather than an item of clothing, and the subsequent regulations that curtail the liberties of his wives. It is because members of Muhammad's generation were still instilled with a loose set of sexual standards that these restrictions are given to protect his wives rather than being motivated by Muhammad's "personal passions, sexual or otherwise."[91] Her primary point is that Muhammad had more or less come to be a "prophet-king" and certain restrictions were needed to protect the honor of his family in the "rough-and-tumble democracy of his day." Would it not have been natural, she argues, for the leader of Islam to adopt practices of the upper classes of surrounding peoples, in particular seclusion of the family and veiling in public? Such an innocent beginning, she asserts, unfortunately "laid the foundation stone of what was to prove in time one of the most stubborn and retrogressive institutions in Islam—the segregation of the women behind curtain and veil."[92] In this scenario politics, not libido, precipitated the descent of the veil.

For Haykel, Orientalists such as Muir "give full vent to their resentment and imagination" over the Zaynab affair and institution of the veil.[93] The problem is that they have chosen fanciful reports and questionable traditions from which Haykel is at pains to distance himself. He begins, however, with a caveat that is more for the Muslim audience than any stray Orientalists who happen to read it. That is, because Muhammad is a prophet, regular rules of social law do not apply just as they did not apply when Moses killed a man or when Jesus was born of an unwed mother. The main thrust of Haykel's argument is that Muhammad "did not marry his wives for lust, desire, or love" even though some Muslim writers have said these things in the past.[94] In the specific case of Zaynab, Haykel appeals to common sense. Here is Muhammad who for almost three decades did not marry a second wife or take a concubine and who had already married a widow who was not attractive. Here is Zaynab who had been brought up in sight of the prophet, who would therefore have been aware of her beauty from the start. The real purpose of the story, in Haykel's mind, is that Muhammad arranged a marriage between his first cousin Zaynab and a former slave Zayd to undo the existing racism of the time against slaves and to reform the practice of letting adopted sons inherit. In this Muhammad becomes an exemplar of obedience to God because he endures social condemnation rather than disobey a divine command meant to place all Muslims on an equal footing. The focus on sex is thus in the mind of the Western beholder, not endemic to a Muslim reading of the story.

What Haykel touts as a scientific approach to Islam's sacred history, Bint al-Shati' regards as a sell-out to Western skepticism. "Haykel seems to forget though," she retorts, "that the story of the Prophet admiring Zaynab, the hair curtain blown aside, the Prophet leaving the house—all this has been written before the world ever heard of the Crusades, by a group of Muslim biographers who can by no means be accused of hating the Prophet or introducing false accusations against Islam."[95] Why, Bint al-Shati' asks, is it so hard to admit that Muhammad could be a passionate lover and why should he not admire the beauty of *Zaynab*? It must be remembered, she continues, that Zaynab's marriage to Muhammad was foreordained by God to serve a purpose. Here she quotes the late twelfth-century commentator al-Zamakhshari that you cannot fault "the aspiration of the heart of man toward some desired object" by reason or religious law "when it does not proceed from the will of man, nor is its presence there by his choice." Like Haykel, she focuses on the lesson in having Zaynab marry the former slave Zayd, who she adds was the next convert to Islam after 'Ali. Unlike Haykel, she believes that Muhammad "felt an overpowering compassion for the young lady forced to marry, in submission to God and himself, someone she did not want."[96] Muhammad's admiration of Zaynab's beauty was normal, but she insists this must be balanced with his self-control and restraint. The prophet fell in love but held his feelings in check; it was God who wedded him to Zaynab for a broader purpose. As for the veil and seclusion, these restrictions were not a punishment on Muhammad's wives but "a symbol of protection, dignity and a desire to rise above the commonplace."[97]

The symbolism of the veil in Islamic texts also intrigues sociologist Mernissi in her discussions of gender. The specific case of Zaynab first appears in *Beyond the Veil* in a section titled "The Prophet's Experience of the Irresistible Attraction of Women."[98] The Muslim male, she asserts earlier, is perceived to be virtually powerless in the face of female sexual aggressiveness. What her Western readers must understand is that the Muslim woman is not frigid in the Freudian sense; she is somehow both *femme fatale* and phallic at the same time. To prove this point Mernissi quotes a few blatantly misogynist traditions that equate the presence of a woman with that of Satan, including the one in which the prophet allegedly said "After my disappearance there will be no greater source of chaos and disorder for my nation than women."[99] Not only is woman "the epitome of the uncontrollable," it would seem that even the prophet of Islam could not withstand the wiles of an aggressive and beautiful member of the opposite sex. On this point, Mernissi complains that Muhammad contradicted his own teaching by telling others to marry for the sake of religion, but in fact marrying for beauty on several occasions. Even when alliance appears to be a motive for marriage, as she admits in the case of the prophet's marriage to Juwariyya, it is feminine charm that Mernissi believes to be the true stimulus for Muhammad. Here then is the dilemma facing Muslims: women are so irresistible that even the best of men is unable to control himself. Hence the veil and seclusion become a lust-based "Islamic" response to male–female dynamics.

Mernissi refers to the prophet's sudden "passion" for Zaynab as "the most significant example of women's irresistible power over the Prophet." As she retells the story, "When he saw Zaynab, who was half-dressed, he felt an irresistible passion for her." Rather than enter, Muhammad "ran off, mumbling prayers," although she adds— wrongly—that we are not given the content of those prayers, as if what he said would not matter. This is critical for a Western reader, who sees the prophet's action as a sign of weakness, whereas most Muslim commentators view it as a sign of self-control and strength.[100] Zaynab told Zayd, who like a dutiful son offers to divorce his wife, which Muhammad at first refuses. "To calm the scandalized clamour of the prophet's contemporaries," Mernissi argues, "the Muslim God made a lasting change in the institution of adoption." She is quick to point out that the prophet was not aggressive or macho in his advances, but rather evinced a basic vulnerability in the presence of such a vivacious female.

Mernissi returns to the Zaynab story later in *The Veil and the Male Elite* to explain the figurative descent of the veil.[101] At the marriage celebration the prophet "was impatient to be alone with his new wife, his cousin Zaynab." A small group of "tactless guests" remained and Muhammad was too polite to tell them to leave. Thus, we are told that the "veil was to be God's answer to a community with boorish manners whose lack of delicacy offended a Prophet whose politeness bordered on timidity." There is little to fault in this analysis, which is based on a close reading of the prophet's biography. But Mernissi argues that such a "minor irritation" can hardly account for such "a draconian decision like that of the *hijab*, which split Muslim space in two?" The broader context, she insists, must take into account the military defeats and doubts undermining the morale of the fledgling Muslim community. Earlier, in her etymological treatment of the term *hijab*, Mernissi concluded that this concept cannot be reduced to "a scrap of cloth that men have imposed on women to veil them when they go into the street."[102] At least, that could not be what it meant at the time of the prophet. In this she may be right, but there is no way of empirically validating sacred history. Her argument is a progressive step for what the future of Islamic views on the gender may take, but a moot point for interpreting past history of Muslim teaching and behavior.

In reading through each of the above rereadings of the Zaynab story a number of rhetorical commonalities stand out. First and foremost, all of the accounts cited above consider this story to be of critical significance for understanding Muhammad's views on gender. Yet, it is more likely the Western prejudice against Islam as such that motivates most of the attention given to the alleged romance of Muhammad and Zaynab. The story itself says little about how females should behave, although it is indicated in the commentaries that Zaynab was so full of herself that she hardly qualifies as an appropriate role model. A second underlying motivation in all of the texts analyzed here is the attempt to represent the sacred history of Muhammad as literal. It is striking that all of the perspectives represented above are passionately argued as if the historical truth of Muhammad's recorded life really mattered. For Muir a story like that of Muhammad and Zaynab is a deliberate distortion of history that he as a trained scholar outside the religion can rectify with critical analysis. The popularity of Muir's text may have been due in part to the emerging context of biblical studies in the latter half of the nineteenth century, when a critical academic approach to religious texts was becoming paramount. It was not that religious narrative was considered by its nature ahistorical, but that it was subject to redaction and alteration, annoyances that make the work of the historian in reconstructing the historicity all the more difficult. Abbott is less concerned with historicity, but still writes as though a discriminating historian can separate the original text from later emendations. Both Haykel and Bint al-Shati' accept the Quran as true revelation, although the former attempts to shift attention away from the legacy of superstitious and magical interpretations. Likewise Mernissi, both in *Beyond the Veil* and more specially in *The Veil and the Male Elite*, accepts the accounts recorded in the authorities she cites as attempts at history. For her, in the long run, it is a conspiratorial male elite that has manipulated and distorted the texts about the "Prophet-lover"[103] whom she is then free to lovingly remake in her own desired image.

Discursive Links

The sword of Muhammad, and the Koran, are the most stubborn enemies of Civilisation, Liberty, and Truth which the world has yet known.[104]

William Muir

One wonders if a desegregated society, where formerly secluded women have equal rights not only economically but sexually, would be an authentic Muslim society.[105]

Fatima Mernissi

It might seem odd to compare the rhetoric of Orientalists, believers, and feminist anthropologists on the single issue of Zaynab as a representation of the gender ideology attributed to the prophet Muhammad. Certain aspects of the approaches discussed are predictable: the old-style conservative Orientalism portrays Islam as inferior to Christianity, Muslim apologists find their religion and prophet to be above reproach, and anthropologists compare the textual tradition with their knowledge and understanding of a contemporary Muslim society. What is of interest to me is not simply what individuals have said about Muhammad's view of gender but rather how this is framed and to what purpose it is put in a broader argument. Given that Muhammad is perceived as the ultimate authority and role model for Muslims, a concern with how Muhammad interacted with his own wives, or women in general, is only to be expected. None of the authors cited above set out to focus exclusively on Muhammad's relations with women, yet in each case it was important to address this issue as integral to the evolving discourse.

Each of the texts contributes to a gendering of Islam, evoking a sense of what it means to be male and female if you were to be a Muslim. Non-Muslim scholars today would most likely cringe at the obvious Victorian overtones of Muir's moralistic pronouncements, but might find in Mernissi's discussion a comfortable and recognizable argument. The non-Muslim might also be surprised that any Western scholar would question the commonly held wisdom that Islam is a misogynist worldview, at least that Islam, which is ethnographically present, forged through the past by a veritable patriarchal male elite. Muslims, by and large, reject the outmoded Orientalist scholarship of Muir and contorted critique by Mernissi as anti-Islamic, and with good reason.[106] Perhaps some Muslims reading my narrative suspect it is yet another Western-fetishist focus on Islam's view of women; it is and it isn't quite that.

It is obvious that the texts analyzed here set out alternative, at times diametrically opposed, points of view. This does not mean that we have to choose one or the other, nor that we should combine the best points for a compromised gender model. Rather, a probing of the ways in which individual authors use the same rhetorical techniques to argue opposing points of view clarifies some of the ongoing shortcomings in the continuing debate over gender and Islam. Valuable and responsible analyses of gender and Islam are available. But as long as my own university library includes both the works of Muir and Mernissi, which might theoretically be read by any student or colleague, it is worthwhile considering how similar these radically different authors are in their rhetoric. And, as long as even a few

ethnographers (e.g., Combs-Schilling) rely on outdated and secondary references to justify simplistic and self-servingly academic versions
of Islam's sacred history, it is important to stress the need for critical
scholarship across disciplines.

Muir and Mernissi, though obviously separated by cultural background, religion, discipline, and gender, both pursue an attack on an
Islamic tradition they allege to be distorted and ultimately immoral.
Muir measures the prophet in light of his own Christian background;
Mernissi attacks those male commentators who she believes fundamentally altered the prophet's real vision of gender. Both see the idealized Muslim male as archetypically a sex fetishist or fiend, evidenced
in large part by his acceptance and alleged promotion of polygyny.
The insistence that a man would need more than one legitimate wife
to satisfy his libido is as distasteful to the nineteenth-century Orientalist
as to the twentieth-century feminist. In neither case is the cultural
milieu of seventh-century Arabia considered apart from the Whiggish
hindsight judgment of the author's present. It is interesting to note
that Abbott and Bint al-Shati' both chide this type of ethnocentric
condemnation of polygyny as hypocritical. What of the Mormons,
asks Abbott;[107] what of Western monogamy, which is often little more
than modern slavery and at inducement at times to prostitution,
queries Bint al-Shati'?[108]

Muir, although he considered himself to be a critical historian,
accepted with little or no skepticism statements and stories that
Muslim commentators have debated quite avidly. So does Mernissi,
particularly in her narrow focus on one particular text on marriage by
al-Ghazali. She styles this single text as a "brilliantly articulated"
Muslim theory of gender, an ideal type to be understood in relation
not to other Muslim scholars but rather to, of all people, Sigmund
Freud.[109] No attempt is made to contextualize al-Ghazali's text, nor
to determine what its various readers thought of it. Nor does it seem
to matter that this text is some nine centuries old, while Freud stands
firmly, if no longer squarely, in the modern world. The absurdity of
such a comparison may not be readily apparent on first reading, since
the Western reader already has a general idea of what Freud is saying
but knows nothing about al-Ghazali. There is an element of irony in
Mernissi's choice, since most feminist criticism of Freudian psycho-
analysis challenge simplistic and ethnocentric conceptualization of
human sexual drives. If a "medieval" imam lashes the feminine to the
Satanic, Mernissi assumes this must be a generic Muslim perspective
and it must be still in force.[110] In fact, the second-century C.E. church
father Tertullian had preempted al-Ghazali by several centuries in
defining woman as the "Devil's gateway" well before Muhammad

received the message of Islam.[111] Should this statement of an early Church father prove a single, misogynist, Christian view of gender that can be cited to explain contemporary behavior of people who call themselves Christians?

Another shared fallacy of Muir and Mernissi is the use of contemporary information to characterize the past. In a footnote discussing the loose morals exhibited in the Quran, Muir informs the reader that the Western traveler Burckhardt met a forty-five-year-old Arab man who had sired children by fifty wives, and that a certain Islamic princess admitted that most women of Mecca have had at least ten husbands.[112] This contemporary anecdotal evidence was cited by Muir to prove the "laxity of morals prevailing" at the time of Muhammad. At a time when few challenged the concept of an unchanging East from the era of the great patriarch Abraham, such sleight of time was common in Orientalist and Biblicist rhetoric. Sir Richard Burton made his reputation, such as it was, doing this in his translation of the *Arabian Nights*. Mernissi pursues the same mode of fallacious reasoning, mixing medieval commentary with Moroccan folklore[113] and juxtaposing statements of the prophet with the 1958 Moroccan Code.[114] Ironically, the first chapter of *The Veil and the Male Elite* is an excursus on "The Muslim and Time," where Muslims are said to suffer from a *mal du présent* assuaged only by a hegira-like flight into the idealized past.[115] In Mernissi's view Western imperialism is to blame for the current identity problem in the Muslim world, since it "obliges all other cultures to fall into line with its rhythm."[116] Political correctness aside, in this case it is Mernissi's own ignoring of time and historical context that allows her to construct a unified Muslim view of gender seemingly outside of time or space and to let her compelling rhetorical style transport the reader there.[117]

When I read Muir more than a century after his first edition, I recognize the ethnocentrism of a Victorian Christian intent on wordhandling the founder of Islam. I would no more send someone interested in Islamic studies to Muir's text in order to learn about the life of Muhammad than I would send a biology student to examine a mid-nineteenth-century text on race. This is not to belabor the fact that past authorities were often representative of the biases of their era, but to assert that our knowledge has mercifully improved. Unfortunately, I find a similar level of distortion in Mernissi's *Behind the Veil* and Combs-Schilling's *Sacred Performances* as that which is so widely associated with Orientalism. To speak of a specifically "Muslim" concept of sexuality or gender, as Mernissi explicitly does, is hardly more justifiable than reifying a changeless "Orient." To ask a question as overtly polemical as "Is Islam opposed to women's rights?,"[118] no matter what answer is anticipated, is fundamentally and prejudicially flawed from the start. If Muir's "Orient" exists only through his will to define and dominate it, then Mernissi's "Islam" and "Muslim Society" are likewise creations of a certain will to control and redefine. Her "Muslim women" and "Muslim men" exist only in the abstract; flesh-and-blood Muslims inevitably differ culturally and individually, as ethnographies have shown time and time again. As Katherine Bullock observes, in Mernissi's vision there is no room for women who choose to wear the veil since "covered women are silent, denied agency, and treated as passive victims of men."[119] Perhaps Mernissi or Combs-Schilling can come close to defining a Moroccan everywoman or everyman, ignoring such annoying social facts as class and status and avoiding comparative ethnographic analysis. Such a construct, no matter how elegant, must not be applied cross-culturally wherever identity is labeled in some sense "Muslim."

Islam cannot be definitively gendered through reductionist rhetoric any more than Christianity, Judaism, Hinduism, or any historical faith that has evolved across cultures and over a large expanse of historical time. It is dangerous to abstract a single gender ideology for even a somewhat discrete cultural group, let alone a large religious tradition relevant to almost one-fifth of the world's population. It is inappropriate to patch together selected traditions and stories recorded over the centuries by diverse Islamic authors into a meaningful reconstruction of what the prophet of Islam really thought about gender or as a foundation for explaining contemporary ethnographic data. The Moroccan women interviewed by Mernissi yield views of gender about themselves, just as the women in a Yemeni village allow an ethnographer there to construct ideas about their gender roles.[120] The fact that both are Muslim makes comparison possible, but it hardly justifies one localized setting indiscriminately serving as a microcosm for the whole. There are well-reasoned examples of how "Islam" is defined within specific cultural contexts rather than Mernissi's approach of a pervasive "Islamic" gender overriding local culture. For example, in her ethnographic analysis of temporary marriage contracts in Iran, Shahla Haeri projects an ambivalent double image of women.[121] Similarly, Jon Anderson argues that Pakhtun in Afghanistan do "not" talk about "different natures for men and women or that women are constitutionally inferior to men."[122] It is the lack of a comparative ethnographic basis that severely weakens the arguments of Mernissi and Combs-Schilling about why Muslims act the way they do.

But why would an anthropologist or sociologist want to speak for as amorphous and idealized a grouping as Islam? As the late

Abdel Hamid el-Zein argued eloquently almost two decades ago, such a monumental "Islam" does not exist, certainly not for the anthropologist.[123] We are better off, he argues, to focus our skills on the many "islams" we can approach anthropologically. Perhaps Mernissi should be taken at her word when she insists she is not trying to "treat Islam and women from a factual point of view," nor is she writing history as "the official narrative that is pressed between covers of gold and trotted out for ritual ceremonies of self-congratulation."[124] Her discourse is ultimately a forum for probing her own identity, arguing passionately for an Islam that she as a modern woman "risen from the harem" can live with.[125] Such a process is important and should not be belittled, given the misogynist views that are well documented. Nonetheless, this is advocacy rather than analysis, didactic rather than dissective. Mernissi believes that Islam, as intended by Allah and superintended by the prophet, was originally and ideally a message promoting gender equality. If contemporary Islamic practice is by and large a patriarchal distortion of this ideal *ummah*, most Muslims would probably demand more than the claim of male bias to explain why Allah would allow His message to be hijacked by so many generations of the devout.[126]

Like Haykel the rationalist, Mernissi the feminist speaks within an avowedly Western frame of thinking to reform an Islam that comes across as too exotic in either a colonial or postcolonial context. Both would sweep away the superstitious, because both need an Islam palatable to their own foothold in a Westernized society. Mernissi invites the reader to "raise the sails and lift the veils" on a dreamlike voyage to what Islam was before a male elite perverted its unique sense of equality.[127] But that dream world of Islam is not grounded outside a textualized ideal. For Mernissi, like Abbott, there is an attraction to the strong and spirited women who surrounded the prophet, especially Aisha as a liberating model for Muslim women. The hope of both Mernissi and Abbott was that their studies would assist the movements toward woman's progress East and West.[128] But how does anyone come out of the dream palace and make a difference in the observable world only in the abstract?

Muir, the Orientalist biographer of Muhammad, was a man speaking primarily to men, his fellow English-speaking scholars. Muhammad was weighed and found wanton by the canons of what an upper-crust Victorian gentleman should and should not do. Mernissi, as feminist, is a woman speaking for women. In her visionary rhetoric, Muhammad is resurrected as an ideal for what the male gender role could be for Muslims. This distinction explains in part why Mernissi's construction of a gender model for Islam has gone relatively unchallenged in print since its first publication in the mid-1970s. Many Western feminists have praised the courageous effort of a Muslim feminist to speak out against the misogynist elements in Islamic tradition.[129] My concern is not to deny such elements, which are hardly unique to Muslim societies, but to question the rhetoric used to essentialize a specific Islamic model of gender taken out of obviously biased source material.

Key to both the old Orientalist view and some feminist manufacturing of a distinct Islamic gender model is the assumption that the critical tradition that has evolved in a Western ideological context is valid for all cultural contexts. Anthropologists have recently provided counters to ethnocentric feminist criticism of non-Western ideologies of gender; this has occurred both in the discipline at large[130] and more recently for the Middle East.[131] As historian Judith Tucker observes, much misunderstanding has resulted from "our assumption that the symbols and content of women's oppression are constant across cultures," and that "the issues of women's liberation as they developed historically in the West should prove to be the same in the Arab World."[132] In reviewing *Beyond the Veil*, Nancy Tapper suggested that this stereotyping of a patriarchal Islam downtreading women can also be a form of self-legitimation by Western women since "their own domestic subordination seems insignificant by contrast, while at the same time their achievements in the public arena seem more significant than perhaps they are."[133] It is perhaps comforting to think that other societies are far more infected by patriarchy. At times it has led to a reformist zeal that non-Western feminists find maternalistic or worse. Sachiko Murata thinks that some feminist critics of Islam advocate an ideal of reform that is "of the same lineage" as Western Christian missionaries.[134] At issue is not a failure of feminist scholarship, which has reinvigorated the analysis of Western stereotyping, but the tendency to essentialize the diversity of ways in which one lives as a Muslim man or woman.

The Orientalists, believers and feminists examined here all take the issue of Islam and gender seriously, at times far too zealously. To play out the metaphor presented at the start of this study, each author cited above is rhetorically at play in the bed of the prophet. Orientalists play fast and loose with embellished stories of Muhammad's domestic life; believers rush to defend the paradigmatic founder of their faith. In an ironic sense, nothing about Muhammad's life is sacred in this discourse, regardless of whether the author is non-Muslim or Muslim. Even the most out-of-character statement, such as Muhammad allegedly singling out women as the greatest threat to a Muslim male, is taken as hard evidence for an endemic sexism in Islam. What is unfortunate is precisely this focus on Muhammad's married life and sayings about women as an assumed paradigm for gendering Islam in the aggregate. It is not surprising that this should be the case among believers, but it is equally the case among critics, be they nineteenth-century Orientalist apologists or twentieth-century feminist scholars. Such a focus obscures a critical appreciation of the elaboration of Muhammad as founder of a major religion. To the extent anyone judges this essentially sacred history by the canons of literal history and reifies a gender ideology on the basis of ambiguous texts alone, the understanding of male–female dynamics in any Muslim community is impoverished.

We can hardly hope to probe "beyond the veil," when it is our own rhetoric that segregates us from the object of our study. If we derogate the missionary zeal of Orientalist Muir, as we must, must we not also descry the disguised didactic bias and shoddy textual scholarship of contemporary writers who, in a sense, disengender Muslims with similar rhetorical arguments? Dredging up the variant ethnocentrisms surrounding the issue of Muhammad and gender, as I have done here, may hopefully further an awareness of the embedding of our own ethnocentrisms.[135] If nothing else, it may bring to the fore fair play in future discussion of gender in Muslim societies.

Chapter 4

Akbar Ahmed: Discovering Islam Inside Out

Being Muslim allows me special insights but also places certain constraints upon me.[1]

Akbar Ahmed

At this point the theologian takes over from the anthropologist.[2]

E. E. Evans-Pritchard

In writing *Nuer Religion*, one of the classic ethnographic accounts of an indigenous religious system, E. E. Evans-Pritchard concluded that he finally reached a point where anthropology stopped and theology took over. Like most anthropologists, he looked at the other's religion as a conscious outsider, neither intending to become Nuer in belief nor to compare it truthwise to a predefined true religion of his own. Ironically, he first encountered the animist Nuer of Sudan as a skeptic of his own religion but was so impressed by the power of their ritual that he later converted to Catholicism. Like Evans-Pritchard, most ethnographers working in Islamic societies focus on what Muslims do and say, the cultural ramifications of doctrine, rather than engaging in theological debate or apologetic polemic. No modern ethnographers have been missionaries intent on converting Muslims, but several have been Muslim themselves.

In *Discovering Islam*, British-trained social anthropologist Akbar Ahmed makes sense of the history and social context of his own religion. Ahmed writes from the conscious bias of a South Asian, extolling the virtues of Pakistani and Indian Muslims from religious icons such as Mawlana Abdul Mawdudi to Urdu poets, a Nobel prize winner, and even the world's fastest bowler in cricket.[3] As an insider Ahmed certainly has insights not available to a nonbeliever, but at the same time, as he recognizes, certain constraints are imposed. The chief constraint, hardly a novel one, is how to objectively analyze a religion that you accept *a priori* as the true religion. How can a Muslim combine anthropology, which claims to be objective, with theology, which is anything but? Alongside this is the cultural trajectory of the ethnographer himself or herself. How can Ahmed analyze Islam in the abstract across geographical space when his own experience and proclivity point to a certain South Asian version?

The question is not simply whether or not a Muslim can represent his own faith objectively as an anthropological exercise, but if religious faith has a tendency to impinge on scholarship that presumes, on some level, to be scientific. The paradox—for many an oxymoron—of faith-based ethnography has not received much attention in mainstream anthropological journals. It is assumed that anthropology requires setting aside personal beliefs rather than trying to interpret according to a preconceived dogmatic frame. In the late nineteenth century, for example, Christian biologists and geologists needed to transcend the established biblical scenarios of a young earth and the literal dust-bowl creation of Adam and Eve to read the evidence for the evolutionary history of the earth and the human species. To a certain extent ethnographers are faced with a similar scenario in which there often is incontrovertible material evidence to falsify specific religious or spiritual claims. Traveling from the secularized West to a "primitive" society has privileged the sense that the religions being studied are behavioral and symbolic systems rather than accurate scientific and historical renderings of a shared "reality." Few anthropologists have in fact been tempted to convert to the "native" religions they study; after all, Evans-Pritchard chose the donning of an Oxford don's wardrobe rather than sacrificing oxen, or cucumber substitutes, on the doorstep of his office. The religion of the other, especially an exoticized other, is something to be explained rather than embraced.

Because Islam is a monotheism closely related to Judaism and Christianity, it is not surprising that an ethnographer might convert to Islam as a result of personal conviction. Bill Young, for example, decided before arriving in Sudan for fieldwork among nomads that he would convert to Islam, with mixed feelings of fascination for the faith and an awareness that being a Muslim would help establish rapport.[4] As some of the Sudanese that he met marveled at the conversion of an American Christian, Young was confronted with a sense of guilt. "I was not certain that I would remain committed to Islam," he reflects, "and was not even sure what that might entail." The questions that he had about Islam went largely unanswered while in the field, so he practiced the faith by copying the rituals of the Rashaayda nomads among whom he lived. In the spirit of

Evans-Pritchard, who also studied Sudanese people, the ethnographer found prayer to be both "a relief and a source of inspiration." As his Muslim identity evolved in the field, he imagined that Islam would remain his faith on return to the states. But his "contradictory feelings" about Islam were not resolved at the time of writing his major ethnographic monograph more than a decade and a half later.

Muslims who have been trained as anthropologists appear to have little problem in adopting a Western secular approach in harmony with their personal faith.[5] This is not surprising, since the idea of undertaking anthropological training is hardly to be confused with entering a seminary or a *madrasa*. Many Muslim anthropologists, such as Sayed el-Aswad, Fadwa El Guindi, and Abdel Hamid el-Zein, carefully separate their faith from their fieldwork accounts. A notable exception is Akbar Ahmed, whose *Discovering Islam* is "part auto-biography, part history, part literature and part science."[6] Beyond his own sense of the faith, Ahmed has attempted in several forums to construct an "Islamic anthropology" that remains within the scientific framework established for the discipline. My concern is not with the truth of Ahmed's approach, any more than it is with the authenticity of an anthropologist converting to Islam. The personal is relevant, but it can seldom be properly understood from the outside. My focus here is on the representation of Islam by Ahmed as an avowedly Muslim anthropologist. It is the logic, not the ontologic, of Ahmed's rhetoric that needs to be addressed.

Discovering Islam

Writing in the tradition of the great Arab historian Ibn Khaldun, Akbar S. Ahmed provides an explanation of Muslim history and society of interest to Muslims and non-Muslims alike.[7]

In prefacing his effort to make sense of "Muslim History and Society," Akbar Ahmed recounts the extreme responses to an earlier presentation of his ideas in Pakistan in which some found him too Islamic and others not Muslim enough.[8] True to his Anglican-made English education, Ahmed chooses to follow Islam as the "middle path," which he sees as an echo of Ayatollah Khomeini's *nah sharq nah gharb*, "neither East nor West." For an anthropologist from the East and trained in the West, this is a suspicious, rather than auspicious, beginning. I doubt Ahmed seriously thinks the spiritual leader of the Islamic revolution in Iran had in fact transcended cultural politics, evaded his Iranian roots or come to terms with Western intrusion.[9] *Discovering Islam* was written, according to the author, in the East, while he was posted as a political officer in a town without electricity, libraries, or academic colleagues. Symbolically he was as far from the West as can be imagined. "More than once," the author informs the reader, "I left off in mid-sentence to settle tribal disputes, supervise flood relief, and attend to visiting VIPs; and on occasion, in the summer of 1986, to cross into the Sind in hot pursuit of dacoits to retrieve a Baluch chieftain who had been kidnapped by them."[10] The resulting book, a mishmash mislabeled a theory of Islamic history and a mixed pickling of contemporary political and social issues, reflects such inflicted inattention to narrative flow and sustained argumentation. Unfortunately, the *via media* of his method ends up in a muddle rather than a viably nongeographical middle.[11]

Let us begin with the author as he positions himself for discovering Islam. His primary authority, at the time of writing, stems from being Commissioner of Sibi Division in Baluchistan. As a government official of Pakistan, it is hard to see how Ahmed avoids being either East or West. It would seem that the "West" in him—the part writing a book for an English audience—is not only in the "East" but is there for a politicized version of participant observation that would raise ethical concerns were he a European serving in the same kind of government post. Settling tribal disputes and setting off in hot pursuit of kidnappers are not generally recommended modes of the ethnographic method. And why are we informed that the book was written under hardship, when it is this kind of difficulty over "there" that foreigners almost always complain about when living in Pakistan? I suggest that Ahmed enacts Khomeini's imagined limbo inside out, unable to escape being in the West and the East at the same time.

What are the author's credentials for writing this book? Obviously, as a Muslim, he is entitled to reflect on his own faith. But Ahmed's expertise, honed in his native Pakistan before becoming a commissioner, lies in ethnography, reflected in two earlier monographs on Pathan and Pukhtun social organization.[12] *Discovering Islam* is not an ethnography as such; it comprises Ahmed's derivative account of the long history of Islam culminating with the interaction of East and West in the postcolonial era. Thus, like Geertz, Gellner, and Mernissi, he views "Islam" not from the field but rather far afield, relying primarily on secondary sources rather than seminal Islamic historical treatises. The back cover lauds the author as "an internationally known social scientist," a mundane statement of fact that few would gainsay, but it is unclear why this should lead to an "objective picture which emerges brilliantly . . ." Most of the first part of the book refers to events that took place long before he was born, so it would seem that a trained historian might have more to say than an anthropologist without access to a library.[13]

The author inhabits his text from the opening sentence of the preface to the generalized "we" of the concluding paragraph. In the first chapter Ahmed's "I" saw "kill a Muslim for Christmas" in a London underground graffiti. This is followed, almost immediately, by a "we" that positions the author with other Muslims. "For Muslims therefore," argues Ahmed, "it is a good time to pause, to reflect, and to attempt to re-locate the main features of, to rediscover, Islam."[14] But the "we" that then takes stock fades imperceptibly into a "we" that needs answers, rhetorically seducing the reader when asking the question "Can we make sense of Muslim history?"[15] The reader patient enough to reach the final section of the last chapter will find an audience-inclusive "We" starting the first four of the six paragraphs; the last one beginning with "Ours." The autobiographical "I" appears anecdotally throughout the book, including a rhetorical presence in a youthful poem that was not liked by his father. These reflections are at times fascinating and penetrating, as they reflect the lived experience of late colonial British rule in India, but they have more to do with an author discovering himself, as he freely admits, than helping a reader discover Islam.

Ahmed's process of discovery is not as objective as the book cover would suggest. "As a Muslim," admits the author, "I cannot write this book as a neutral spectator or observer. I am also a participant, an actor in the drama."[16] The sentiment is sincere, but the consequences are not probed. Participant observation is here turned on end into an observant participation in which the author is able to see "some of the majesty of Islam." The actor not only plays a part, but wishes to live it off the stage as well. Ahmed implies that to be "neutral" means a lack of attachment, as if a secular anthropology demands this. He fails to note that ethnographers often develop a strong emotional bond with the people they study. For example, the anthropologist Stuart Schlegel writes an eloquent tribute to the Filipino rainforest people he studied but who were later brutally murdered in a political ambush. Schlegel is anything but neutral in his feelings, even though he is not describing his own religious faith.[17]

The issue is less neutrality than objectivity, although Ahmed tends to conflate the two in his prose.[18] Eschewing the postmodern reflexivist critique of ethnographic objectivity, the fact that he is both actor and observer is cited as a rationale for presenting "an accurate, objective view" simply by positing a balancing act of the two roles. Such balance, for Ahmed, is achieved by not degenerating into "aggressive polemics or apologetics."[19] Thus, the criterion for objectivity becomes how aggressively the argument is posed rather than the accuracy of what is being said or the relevance of parts in explaining the whole. Ironically, given the earlier citation of Khomeini, it seems to be his "encounter with the West" that allows the author to view Islam, at least theoretically, as part of a universal civilization.

Like Geertz and Gellner, Ahmed begins his discovery with a self-serviced warning. "History will be presented here in broad sweeps, in ways which traditional historians may not approve. Society will be generalized about in a manner calculated to cause anguish to traditional anthropologists."[20] True, on both accounts. But who are these "traditional" historians and anthropologists, who come across in the rhetoric here as establishment, stuck-in-their-Ivory-Tower guardians of the past? Is an academically defined "tradition" always so suspect that it must be subverted? Perhaps scholarly disapproval stems from Ahmed's failure to follow sound historiographic methods and his insistence that merely relating case studies passes for credible social science. Simplifying arguments and drawing broad conclusions are the short cuts of journalists, not suitable goals for competent scholarship. Certainly there should be cause for concern over Ahmed's own approach as a Muslim anthropologist if we are to take seriously the claim that the "modern Muslim intellectual exists in a state of despair, torn between an ideal world he cannot order and a reality he cannot master."[21] It is hard to see how this context allows for adopting a middle way or a balance and at the same time avoiding the East–West cultural divide that permeates Ahmed's text from start to finish.

One measure of the failure of the author to transcend the Orientalist grip he derides can be found in the book's appendix on Muslim chronology. The whole time frame is oriented to B.C. and A.D., not even a nod to the de-Latinized C.E. The two items mentioned for B.C. are thoroughly Orientalist in the Saidian sense. Beginning a specific "Muslim chronology" with the "First mention of Arabs, in an inscription of Shalmaneser III" (853 B.C.) may be an ancient ethnic marker, but being Arab in Old Testament times hardly explains what it means to be a Muslim. Similarly the "Expedition of Aelius Gallus to southern Arabia" (25–24 B.C.) suggests that Muslim history is activated by what a Roman legion did. What does this inconclusive historical event even remotely have to do with events leading up to Muhammad? The Islamic era is noted as beginning A.D. 622, when the Prophet was forced to emigrate from Mecca to Madina. The Battle of Badr, two years later, is heralded as the "first major Muslim victory," but Ahmed ignores the critical defeat of the Muslim forces a year after that at Uhud. To say that Muhammad conquers Mecca in A.D. 630 is convenient shorthand, but is it not worth noting that this was done without bloodshed? The other listed events in caliphate and colonial history of the Middle East, with an occasional tidbit on Pakistan, emphasize who was assassinated, murdered, conquered, captured, revolted against, sacked, defeated, occupied, and the like. This is an odd listing for an author who is advertised by the publisher as challenging the Western bias that Islam is aggressive and fanatic. Indeed, the abrupt ending in 1986 with "Martial Law lifted in Pakistan" and "Professor Ismail Faruqi and wife murdered in the USA" sounds an ominous counterpoint to a book that is advocating the building of bridges between peoples.[22]

Ahmed's discovery approach consciously idealizes Islam. Specifically, this is the "ideal or pure type" advocated as a sociological model by Max Weber, although the reader is dutifully warned that such an averaged ideal is an approximation rather than a substitute for reality. Weber was concerned primarily with religion as "a particular type of social behavior" rather than defining the "essence" of any given religion,[23] but the ideal resorted to by Ahmed is *tout court* the acknowledged essence of Islam: the "supporting and inter-locking" elements of the Quran and the prophet Muhammad. This is indeed the well-known ideal in which Quran and *sunna*, the revelation and the life of the prophet, lead to the *shariah* or path for Muslims. "They provide us with a good idea of how a person ought to conduct himself to be called a Muslim," notes Ahmed. "The ideal aims at paradise in the next world and satisfaction, if not success, in this one. We thus have not only a way of looking at the world but of living in it."[24] This is what Muslims say they believe, a starting point rather than a novel piece of information or critical analysis. The Quran and information about Muhammad are textual sources that must be interpreted; it is the way in which the texts are acted upon that must inform a sociological or anthropological model of the religion. For a Muslim, Ahmed's "ideal" may indeed be held to be "eternal and consistent," but that is a statement of belief, not an explication of the behavior of believers. None of this is Weberian, whose ideal types explained, rightly or wrongly, the function of religious behavior—the role of a charismatic prophet, for example. Ahmed fluctuates indiscriminately between the idea or "vision of the ideal" as a notion that Muslims have in their heads and the presumed sociological usage of ideal-type as a model for explaining behavior.

Ahmed's discovering falters even before a specific argument is made. He assumes something called "Islam" as an independent force entering history. "Islam comes with definite, specific ideas and does not encourage duality," he proposes.[25] Islam is thus for Ahmed the "Great Tradition of a world religion" forced to encounter the "Little Tradition of local, regional, village culture." Is this Ahmed the anthropologist speaking? Apparently he had not read what his anthropological colleague Abdul Hamid el-Zein was saying a decade before his writing.[26] Can we call any set of statements about the ideal in a religion "definite" and "specific" simply because that is what they are believed to be? Where does the "Islam" he valorizes come from? If it is indeed the universal religion that supersedes all other religions and if the Quran is a revelation unlike any other, then the only conclusion that can be made is that Ahmed discovers Islam from a prior sense of faith. If the ideas of Islam are definite then they must exist outside human appropriation of these ideas in actual societies; the history of interpretation suggests that many of the ideals in Islamic dogma and practice have been anything but definite and fixed. And what does it mean to say that Islam does not encourage duality? Are Quranic concepts dividing good from evil, believer from unbeliever, heaven from hell, meant to discourage Muslims from thinking that the world can be approached in binaries? The ideals and ideas discussed by Ahmed may or may not be true, but deciding that truth is a theological exercise only.

When a theory of Islamic history is finally proposed, Ahmed attempts to return to sociological ground by arguing, "there is a dynamic relationship between society and the striving of holy and learned Muslims for the ideal."[27] The two "traditional" theories rejected quite rightly by Ahmed as failed explanations are the cyclical (rise, fall, rise) model fleshed out of Ibn Khaldun's *Muqaddima* and the decline-and-fall theory attributed to Eurocentric historians like Oswald Spengler, Arnold Toynbee and Bernard Lewis, as well as, ironically, Ayatollah Khomeini. These are straw theories for Ahmed, as indeed for a range of competent scholars of Islamic history. Ahmed wishes to plot a course between an Arabocentric Tunisian scholar and Arabophobic apologist historians for Western civilization. Islam is not "rise and fall" akin to the Roman Empire, but a perpetual cycle of risings, fallings, and new risings. "Instead we discover a rhythm, a flux and reflux, a rise and fall, peaks and troughs in Muslim society attempting to live by the Islamic ideal."[28] I am tempted to call this a circular argument, but in fact it is simply a flowchart with feedback loops.

Ahmed is right to criticize the ethnocentric assumptions and methodological limitations of the linear reduction of Islam to a timeline determined by who won the battles, even though this is precisely the timeline he appends to his text. Muslims, in a political sense, come and go, but an ideal of Islam endures to be constantly reinterpreted.

"When the Mongols captured and sacked Baghdad, the capital of the Abbasids, they destroyed not Islam, but a corrupt and decadent society," argues Ahmed.[29] But they also put to death a large number of ordinary Muslims and then promptly adopted Islam as their own. The Ottomans similarly invaded the Muslim heartland and converted. European powers and bad decision-making from the inside took apart the Ottoman caliphate in the early twentieth century. Nationalism and socialism altered the politics of Islamic countries only to result in Islamist unrest and calls for Islamic revolution. This standard emphasis on risings, fallings, and new risings only serves to accentuate the political. This is a poor way to discover the meaning of Islam, which hope fully has been more than an abstract ideal far removed from the behavior of really violent Muslims over the centuries.

Ahmed proposes a hypothesis to go beyond the trite observation that Islam has survived the vicissitudes of history. "The farther from the ideal," he suggests, "the greater the tension in society."[30] A further component of this hypothetical observation is that the rapid expansion of Islam follows a "law of diminishing returns" in which the greater the numbers, the less satisfactory the results. His underlying assumption is that the further from the age of Muhammad, in time and sheer numbers of Muslims, the more disharmony for the Muslim community as a whole. Islamic history is thus read as ultimately in regression, a retrograde big bang theory for its origins rather than a steady state. All of this echoes, for the author, the words of Usman Dan Fodio, who led a jihad against the Hausa in the early nineteenth century: "Islam has been flagrantly abused by corrupt rulers. We must return to the Golden Age."[31] That such sentiments have been articulated by Muslims of various persuasions is common knowledge and hardly unique to Islam. But this is neither a sociological nor an anthropological theory. The behavior of Muslims is influenced by particular interpretations of the Quran and the life of Muhammad. The so-called Golden Age referred to by reformers never existed in real time. What does it matter if Muslims, at any given time, are further away from the start of the faith when the issue should be how they act on beliefs in a given social setting at any time? Islamic beliefs, from an anthropological viewpoint, are not etched in divine stone but dependent on cultural factors and are inevitably in flux. The envied Golden Age, injudiciously admitted into evidence by Ahmed, is a mirage.

Despite its methodological flaws, Ahmed's hypothesis of decline over time can be readily tested from the available evidence in Islamic sources. The events Ahmed finds relevant to include in his rendering of Islamic history suggest that decline set in as soon as the Prophet was fresh in the grave. "Islam's rapid success," notes Ahmed, "carried within it the seeds of crisis."[32] This is somewhat of an understatement. Uthman, the third caliph, was assassinated, as were both his predecessor and successor. "His body lay unburied for three days as the assassins plundered the treasury," adds Ahmed.[33] 'Ali, the Prophet's son-in-law, was murdered by fellow Muslims while praying in a mosque. His son Husayn was martyred at Kerbala. This tale of woe casts a dark cloud over an Islamic community as yet without a definitive written text of the Quran, formal biography of the Prophet, theological system, or legal school. Is Ahmed seriously claiming that Muslim atrocities have somehow become worse over the centuries than this initial civil and ideological *fitna*? Is Bhutto worse than Yazid? The account given by Ahmed does not support the contention that an initial Golden Age has simply dimmed over time, unless we assume that it went dim the moment the Prophet died. That can be assumed, of course, and it is the stuff of theology.

Ahmed relates the story of Muhammad and the early caliphs as though it was established history. A devout Muslim may accept this sacred accounting as an article of faith, but an anthropologist should approach it from the standpoint of relative skepticism in the same way that any set of religious dogma might be evaluated. Ahmed claims to be writing anthropologically, but his rendering is as romantic as it is abbreviated. Concerning *Arabian Nights* vintage caliphs, we read: "In their high, inaccessible palaces the royalty seemed cold and aloof; perhaps they had no hearts underneath the glittering gold and silk they wore."[34] How did the anthropologist a millennium later discover this psychological profile? Ahmed illustrates his description of less than ideal "Muslim rulers" with an extended courtier's anecdote about a caliph's passion for a concubine. Then we are treated to crusade-era Saladin. "Although not an Abbasid he also represents that age. Moreover he represents the Muslim ideal," declares Ahmed.[35] "To the European he came to symbolize ideal, romantic Muslim chivalry." Ironically, Ahmed succumbs to the same sort of legendary hubris in touting Saladin as the benevolent redeemer of Jerusalem. In yet another anecdote, he mentions a French general in 1920 who stopped at Saladin's tomb in Damascus to say "Awake Saladin, we have returned."[36] What some might read as a statement laced with vengeful sarcasm, Ahmed sees as a tribute to the "lustre" of Saladin's name. At any rate, the modern-day Saladin at the time was Prince Faisal, who beat the French general to Damascus—before being asked politely to go back to British-liberated soil.

The Islam that Ahmed discovers is a decidedly *shi'a* version. Consider his discussion of the "ideal caliph" in which only a particular idealized view of 'Ali is presented. We, as readers, are informed that 'Ali was "brave and scholarly" and that the "moral authority of 'Ali, as the rightful caliph, challenged and confronted the material power and wealth of his rival Muawiyah, the governor of Syria."[37] True or not, this is a partisan perspective, not a pan-Islamic ideal. Whether or not 'Ali should have been caliph is a theological issue, hotly debated historically. The politicization of Muslims over this theological issue is the relevant historical, sociological, or anthropological concern. Yet when Ahmed examines the historical feud between *shi'a* and *sunni*, he is at pains to defend a certain set of views. For example, Ahmed makes it clear that Muslim women are better off in a *shi'a* system, as though specific cultural features are immaterial.[38] Missing in this ahistorical discovery channel are the social and economic factors that led to political and domestic differences, both ideal and in the real world?

The fundamental flaw in *Discovering Islam* is one shared by *Islam Observed*, *Muslim Society* and *Beyond the Veil*: the Islam represented is idealized as an essence rather than analyzed as an evolving culture-bound dynamic of belief and behavior. There is ethnography in the text, even though much of it is anecdotal, but Ahmed invariably targets "Islam" as a material object and "Muslims" as specific agents acted upon by a generic stream of Islam. "Ours has been a sociological exercise," Ahmed concludes, "not a theological explanation of Islam." In fact it is neither. A sociological approach, even a poor one, must focus on the function of "religion" in specific social contexts. Citing Weber, Geertz, and Gellner does not make Ahmed's own self-discovery sociological, nor does it provide a perspective by which to understand the behavior of Muslims acting on their ideals. Anthropology, despite Ahmed's attempt to follow a specifically "Islamic anthropology," is not "well placed to define the Muslim ideal and assist in its construction."[39] Only individual Muslims can define such ideals; it is the task of the anthropologist or any objectively inclined scholar to trace the links of such ideals in a given cultural setting. Ahmed comes nowhere near achieving such a task.

Toward Islamic Anthropology

We may define Islamic anthropology loosely as the study of Muslim groups by scholars committed to the universalistic principles of Islam, humanity, knowledge, and respectful tolerance, and relating micro village tribal studies in particular to the larger historical and ideological frames of Islam. Islam is here understood not as theology but sociology.[40]

Akbar Ahmed

What would a specifically "Islamic" anthropology look like? Akbar Ahmed sees it as targeting a specific group, Muslims, and at the same time accepting a commitment to certain "universalistic principles." "The charter of an Islamic anthropology is thus humankind," he concludes.[41] The content part is no different from an anthropology of Islam, applying anthropological methods to the study of Muslims. But the unique and loose twist here is that the anthropologist submits to a set of principles that are assumed to be universal but left undefined. The main signifier is a commitment to "universalistic principles of Islam," which the author clearly sees as compatible with humanity, knowledge, and tolerance. But from where does Ahmed derive these principles if not from theology, from an interpretation of the Quran and life of Muhammad? Who determines which ideals are universal, a rather dubious goal for a discipline that has shown over and over again that there are few, if any, universals across human cultures? Surely there can be principles of humanity, knowledge, and tolerance outside an Islamic perspective. The Islamic anthropology advocated by Ahmed assumes genuine Islam can only be humane, encouraging of knowledge, and tolerant. This can hardly be faulted as a statement of faith, but it fails to account for the obvious fact that many Muslims have not acted humane, encouraged knowledge, or favored tolerance and moderation.

Ahmed is well aware of the problems in such a constricting definition. Does this not open the door to Jewish, Hindu, and Buddhist anthropologies, he queries? "Are we not thereby restricting a universal science to particularistic groups in the face of opposite trends in the discipline?"[42] The advocate for an avowedly "Islamic" brand of anthropology prefers to resolve these "complex questions" with "simple answers." But he merely begs the question by revisiting the canard that anthropology has already been biased through its development during the colonial era. After all, he argues, there are national schools and ideological variants of anthropology, so why should adding an Islamic anthropology to the list "raise so many hackles?" The logic here is indeed simple, but skirts the issue. Social anthropology in Britain may appear to be a distinct approach, but it is by no means homogenous, nor is every British anthropologist an apologist for his ethnicity. Marxist and feminist anthropologists inevitably carry certain presuppositions into their analysis, but regarding their foundational texts as a revelation from God is not, or certainly should not be, part of this. Indeed, there is a good reason why it is hard to find Jewish, Hindu, Buddhist, or even Nuer anthropologies; anthropology has evolved as a secular discipline in light of earlier lessons learned from viewing other cultures through a specific religious ideology.

It is important to note that Ahmed is not simply saying Muslims should do anthropology. "Muslim sociology for Islam," he argues, "is the manifestation of its theology."[43] Islamic endeavor leads to Islamic anthropology, in Ahmed's view, because Islam emphasizes the pursuit of knowledge "even unto China," as noted in a famous *hadith*.[44] "People are exhorted to contemplate, to think of and marvel at, the multitude of variety in the heavens and on earth," preaches Ahmed, who provides a Quranic justification. But here the avowedly Islamic anthropologist misses a crucial point. Muslims are supposed to contemplate God's creation, not doubt what the Quran has said unequivocally about the creation. Anthropology is not about marveling over the world as a manifestation of God's power. The anthropologist is theoretically committed to pursue knowledge even to the Amazon forest, guided in large part by a dogged belief in cultural relativism. No Muslim can comfortably step outside the belief in Allah as creator and still pretend to call what is done "Islamic" unless the word is redefined out of its historical context. An individual Muslim certainly has a right to do so, at least outside those regimes that view this as apostasy, but it will surely be seen as misguided by the vast majority of those who follow the faith as God's truth.

Why should Islamic anthropology be added to a discipline that Ahmed finds rife with equally biased approaches? How could an "Islamic" approach improve such a tainted discipline? Ahmed observes that a Muslim must approach anthropology as an applied science. Indeed, it is his understanding of Islam that a Muslim in any field cannot be passive but "would attempt solutions to the major social problems facing humanity in the late twentieth century." This seems innocuous enough, if perhaps naïve, but the floodgates are let open with the following comment: "The Islamic use of the word *jihad*, striving—popularly seen as religious war—to better the world, is apt."[45] Surely such a term is anything but apt, given both its contradictory theological and political connotations. Muslims exercise a greater *jihad*, as the prophet called it, to gain mastery over the imperfections and evil in their own lives; the lesser *jihad* of fighting to defend the Islamic community is a phenomenon to be studied by anthropologists not a process to be ignored. Islamic teaching would suggest that there is only one way to better the world and that is according to principles laid down in the Quran and spoken about by Muhammad. Ahmed must assume that the pursuit of knowledge, even to China, could not possibily invalidate the basic tenets of the faith. Theology again must trump the agenda.

In the process Ahmed privileges what he calls a "more general, more practical application" over extended fieldwork "with a remote group on an exotic aspect of their lives." This is a classic false dilemma that ignores the complementary role anthropology has played in the social sciences. Ethnography need not be at the edge of nowhere—an edge difficult to find anymore—nor is it true that sound ethnographic methods cannot be applied by concerned scholars. It is absurd to argue that Islamic anthropology would "act as a corrective to the notorious ethnocentricity of much of Western anthropology." Replacing one kind of ethnocentricity with another would only perpetuates bias. As Edward Said noted in *Orientalism*, the answer to the failures in Western Orientalist dogma is not an equally one-sided Occidentalism.[46] The correction to faults in secular anthropology will not come from submission to a particular historical religious tradition.

What would make an overtly Western discipline "Islamic"? Ahmed teases the reader with the spurious claim that the scholar al-Biruni (died 1048 C.E.) "deserves the title of the first anthropologist."[47] It is apparently enough that al-Biruni, a prolific historian and scientist long studied by Western Orientalists, presented his findings with "objectivity and neutrality" and did so cross-culturally. The same argument has been made over and over again for Herodotus, who certainly has seniority. But it is nonsensical to suggest that this Muslim scholar approached his work with the same kind of objective and neutral attitude assumed to be the case in modern anthropology. First of all, if modern-day ethnographers have been excoriated for not recognizing their own ethnocentric biases, how could we expect a scholar from a millennium ago to do so? Second, Ahmed confuses the general respect held for al-Biruni as a scholar discussing technical matters in science with his inevitable ethnic and religious biases. Al-Biruni, not unlike many puffed up icons of academe today, was highly opinionated and brooked little criticism. Consider, for example, his opening salvo in his classic materia medica:

> If someone tries to question the veracity of what I have said, he will have to confess to his ignorance. If afterwards he cultivates the spirit of enquiry and learning, he will be succoured by God. If, on the other hand, he turns his face away and walks homeward strutting on his possession of ignorance, his face will be ignited by the fire of Hell.[48]

What kind of a model for anthropology is al-Biruni when he describes as "inexperienced and foolish" those scholars of his day who questioned the given ages of the antediluvian patriarchs in Genesis?[49]

A major rhetorical problem in Ahmed's call for an Islamic anthropology is the way in which "Islamic" can stand for just about anything. A case in point relates to an article in which Ahmed called for the concept of an "Islamic district" to be applied to the on-the-ground reality of rural Pakistan, where he actually served as a district officer. In critiquing this notion, Tahir 'Ali argued that "district is a construct of the British colonial administrative scheme which, of course, was applied to peoples of diverse religions and cultures."[50] Ahmed's response is quite telling: "If we can categorize the Republic of Pakistan as Islamic, the constitution of Pakistan as Islamic, and despite strong protests by Pakistan, its nuclear programme as Islamic, why can we not categorize the districts of Pakistan as Islamic?"[51] Of course, we *can*, but why should we? If anything that any Muslim does anywhere is equally "Islamic," then the term is stripped of any religious significance. In this sense any Muslim who does anthropology is *ipso facto* doing Islamic anthropology.

A careful reading of Ahmed's writing suggests that it is not the idea of an Islamic anthropology that he is advocating, but rather a specific kind of idealized Islam. "Islam is a universal religion speaking to humanity," he attests; this can only be interpreted as a theological statement. What of those parts of humanity that do not choose to listen, that do not consider the life of one small community in seventh-century Arabia as a paradigm for a Golden Age, that look at the recorded history of Islamic empires and see ample evidence of intolerance and inhumanity? The fact that Islam has often been unfairly represented in Western discourse does not mean that it is thus immune from criticism as a faith with a documented history. When Ahmed avers that "Islamic anthropology is thus ideally placed to fulfill Islamic endeavour," it is important to go beyond what Ahmed or any other Muslim would like Islam to be and focus on how specific interpretations of Islamic ideals influence behavior in the real world.[52] It is certainly appealing that Ahmed proposes his vision of Islam as a charter for humankind, but he does so in admitted opposition to other Muslims who reject anthropology mainly because it is Western.[53]

The context of Akbar Ahmed's suggestions for making anthropology compatible with Islam is a conference held in Pakistan in 1984.[54] The objective of the conference was to develop a specifically Islamic "school of thought" relevant to existing academic disciplines, based on principles "from the values of Islam" and at the same time "worthy of the greatest empiricism."[55] While Ahmed's contribution is an attempt to reorient rather than reject Western anthropology, the overall project for which it was written is quite alarming in scope. In his inaugural address at the conference, Mahathir Muhammad admonishes Muslims to go beyond achieving "absolute command" of Western disciplines in order to "integrate the new knowledge into the corpus of the Islamic legacy by eliminating, amending, reinterpreting and adapting its components according to the world view of Islam and its values."[56] Who decides what is to be eliminated? What are the unique criteria that Muslim scholars would have to amend in order to reinterpret scientific and social theories? Given that there is no single worldview among Muslims, a point made eminently clear by the contributions at the conference, and that there is widespread difference on Islamic values, how could a serious anthropology survive such theological censorship?

In addition to Ahmed's agenda for an Islamic anthropology, the conference featured a critique by Muhammad Ma'ruf of "evolutionism" in Western anthropology. An evolutionary framework is recognized by this Muslim author as "a fundamental idea of the discipline, so much so that if one were to remove its permutations in the various subfields of anthropology, one would be left with a large number of different mini-fields of inquiry without any internal connection."[57] Ma'ruf is right; contemporary anthropology is evolution or it is nothing. There is no easy way to reconcile a religious belief in divine creation, as articulated in the scriptures of the monotheisms, with the guiding principle of the sciences that all forms of life are related and share a universal genetic code. Ahmed and other Western-trained Muslims suggest that Islam is large enough to accommodate Darwin and the Big Bang, but I suspect many Muslim intellectuals and most theologians would not agree. How can anyone remain an anthropologist without looking at all societies as variations of an evolving human species? If any particular religious dogma overrides the accepted findings of scientific inquiry, anthropology becomes a philosophical handmaiden to the theologian. Could it be otherwise in the idealized Islamic framework advocated by Ahmed?

Reaction to Ahmed's call, inclusive as it tries to be, has been limited among anthropologists of Islam, whether Muslim or not. "Islamic anthropology is no more easily dismissed than any other '-ism'"; suggests Richard Tapper, an ethnographer with extensive experience in Iran and Turkey; "it should be taken seriously because it addresses a wide audience, avows its ideological base, and invites critical discussion."[58] The fact that calls for such an ideological variant of the discipline have been addressed almost exclusively to fellow Muslims makes it difficult to evaluate the intentions of Ahmed and other Muslim anthropologists. Ahmed, in particular, describes an outdated version of anthropology, even at the time he was writing. In his account of Islamic anthropology, the theory and methodologies of anthropology are conflated with the idealized statements offered to introductory anthropology students. "Anthropology makes us aware of the essential oneness of man and therefore allows us to appreciate each other," he informs his predominantly Muslim audience. Would that this were true, but here again is an example of substituting what anthropology claims to be about, with how the findings of anthropological research are actually used. From Ahmed's perspective, he could as easily have replaced "anthropology" here with "Islam." In point of fact, that is precisely what his Islamicized anthropology does.

Ahmed moves toward Islamic anthropology by distancing himself from what he claims are the faults of Western anthropology. His critique of Western ethnocentrism parrots reflexivist deconstruction underway from a number of quarters, but without attention to the theoretical models arising out of literary criticism and philosophy.[59] Telling fellow Muslims that Western anthropology is "in a state of general theoretical stagnation" implies that it is ripe for Islamization.[60] The reader is also treated to a brief foray of Orientalism-bashing à la Edward Said in order to further stress the contaminating influence of most non-Muslim interpreters of Islam. Unfortunately, Ahmed spends all but a few pages in his conference article rehashing the rise and assumed free-fall of social anthropology. Beyond this, he is more concerned in showing that Islam has been misrepresented than demonstrating how Islamic values could rectify the stated problems within the field, let alone why a non-Muslim would want to adopt such a sectarian approach.

Ahmed finally attempts to illustrate his argument for an Islamic anthropology with reference to the Prophet Muhammad: "The life of the Prophet (SAAS) needs to be produced in simple and clear terms for the contemporary generation of Muslims. As his life and example remain the primary paradigm of Islamic behavior, the exercise is vital to an understanding of Islam—both for Muslims and non-Muslims."[61] The anthropology part, for Ahmed, is all content: the social roles of Muhammad, his humility, humanity, gentleness, love of children, and kindness to animals. Citing examples from the extensive hagiography of the Islamic prophet, Ahmed concludes that the examples in the sources "speak of a man of extraordinary perception, goodness, and gentleness." As a biographical reconstruction of Muhammad this is all theology without even a pretense of searching for the "historical Muhammad." Anthropology, certainly the British functionalism described at length in the same article, has little relevance to the exegesis of religious texts, unless comparison is supposed to be made between the religious depiction of Muhammad and ethnographic observation of contemporary Arab tribesmen. Such comparison could actually reverse the progress in anthropological theory, reverting to the time when Arabian Bedouins were lauded as the unchanged mirror of life for the biblical patriarchs.[62]

To place this example in context, suppose that an avowedly Christian anthropologist set out to reproduce the life of Jesus. The primary sources would be the gospels, all written and redacted after the fact. Exegetical scholarship can establish what the texts are saying, but cannot prove the historicity of any specific claims made about Jesus. Was Jesus born of a virgin? Devout Christians and Muslims usually say yes; this is what their respective scriptures say. Did Jesus raise a man from the dead? Did Jesus rise from the dead? These are questions to be approached theologically, according to the stock given to the texts. Are the recorded words of Jesus really from a historical Jesus? If theology demands that the holy texts be accepted as literally truthful, then it is hard to see how the methods of modern anthropology have any relevance apart from raising serious doubts. Anthropologists rarely try to make "truth" claims about any particular religion they study, even though it may be clear they do not accept certain beliefs. Similarly, there is little role for an apologetic approach in which an anthropologist would attempt to justify his religious belief. Yet Ahmed wishes to read Islamic sources about Muhammad without questioning the kind of prophet his personal vision, shared by many others, demands. There is nothing wrong with such a belief, nor can it be said that he does not have a grasp of some ultimate truth that anthropologists must inevitably miss. But this is not anthropology.

Ahmed's call for an Islamic anthropology is problematic from the start. "This study is speculative and concerns a difficult and complex subject. Its task is made more difficult as it defends a metaphysical position, advances an ideological argument, and serves a moral cause," he contends.[63] Anthropology as a scientific discipline, despite all its faults, is in no sense a defense of any particular metaphysical position, especially a religion claiming to be the sole legitimate revelation for humanity. If the Muslim anthropologist finds a discrepancy between his belief and the results of anthropological analysis, he must in Ahmed's scheme side with his belief. To do otherwise is to admit that his faith is subservient to the mundane human quest for knowledge. To think that he can under all circumstances remain a devout Muslim and at the same time pursue an objective and empirical investigation of Muslim society as an anthropologist is naive, at best. Good anthropology, observes Richard Tapper, always has "subversive potential."[64] Rather than accepting the inviolability of any given ideology, it constantly challenges what is commonly perceived. In this sense some Muslims are right to be wary of the findings of secular anthropology.

Ahmed assumes, rather than demonstrates to the satisfaction of his peers, that an insider would have a better understanding of Islam.

He is careful not to assert that only a Muslim anthropologist can correctly interpret Islam, but if non-Muslims like Gellner and Geertz are capable of producing "some of the finest material in Islamic anthropology," what purpose can an Islamic anthropology serve except as a forum for Muslims who wish to have a foot in both East and West.[65] Other Islamic scholars in the West have not been so tolerant of critical study of Islam by non-Muslims. Consider, for example, the commentary by Abdul-Rauf to a volume on contemporary approaches in religious studies to Islam:

> Why have certain orientalists wasted so many precious years of their lives trying to reorder the text of the Qur'an chronologically under the assumption that a human hand played a role in the formation of the text? Such programs of research are not merely an offense to the consciences of Muslims, but are also misleading and thus unworthy to be considered as scholarship.[66]

As Brian Turner observes, Ahmed's representation fails because "he does not fully face up to the critical implications of postmodernism for traditional Islam."[67]

It is not that Akbar Ahmed fails to grapple with postmodernism. In *Postmodernism and Islam: Predicament and Promise*, there are references to Adorno, Barthes, Baudrillard, Bourdieu, Derrida, Foucault, Jameson, Lyotard, and Said, all part of the post-canon.[68] More significantly, at least for a general reader, Ahmed sweeps through postmodern pop culture from Marlon Brando to Kevin Costner, from Jimi Hendrix to Billy Joel. Along the way there are guest citations to John Belushi, James Dean, Clint Eastwood, Jane Fonda, Mick Jagger, Spike Lee, John Lennon, Dean Martin, Elvis Presley, Arnold Schwarzenegger, and Andy Warhol. Through his writing Ahmed implies that referencing critical theorists and reviewing pop culture can offer promise for a better understanding of Islam by non-Muslims and Muslims. I suggest that it will take more than writing about postmodern critique to reach the ultimate aim of his book, which is to call for less bashing of Islam and hope for a more tolerant world to rise from the ashes of current secular cynicism. The predicament, which is hardly unique to Muslims, is that the postmodern despairs at meta-theorizing in either the old positivist model of academe or the dogmatic creeds of established religions. Muslims, including Ahmed, can certainly think their way through such temporary cultural criticism, but if Ahmed's idealized model of Islam is not a meta-theory, what is it?

The failure of Akbar Ahmed to articulate a distinctive Islamic anthropology does not mean that either Islam or anthropology is deficient. The problem stems from combining two different approaches to making sense of the world. Different need not mean antithetical. Anthropology should not be accepted as a secular substitute for religion. Ahmed is right to point out that anthropology helps us understand ourselves by understanding other societies, but he idealizes the discipline by claiming that it realistically fosters appreciation for the oneness of humanity. The awareness of oneness and the ability to better appreciate others need not depend on ethnographic research. Reading about the Yanomamo may strike one person as a case for the unity of mankind, but only serve to accentuate cultural differences in the moral judgment of another person. Indeed, over a century of anthropology as a formal science has hardly made the world a better place. The anthropologist provides information through study of human cultures, but interpretation of the humanitarian or spiritual implications of such observation transcends the academic discipline. The anthropologist always operates within a worldview, whether that is secular or religious, conservative or liberal. But worldviews are not easily boxed up in real time. There will always be a point, as Evans-Pritchard knew so well, when anthropology gives way to theology or even philosophy. Ahmed's faith-based anthropology leads to a philosophical point of no return.

Epilogue

Muslims Observed: The Lessons From Anthropology

It is not possible to break ground for new research on the prophetic discourses to be found in the major corpora of the three monotheistic religions without starting with the liberation of thought from all theories, dogmas, and imageries bestowed by the self-founding and self-proclaiming theologies of each community against its rivals. Then and only then can one tackle Islam itself in an inquiry of an anthropological character on the emergence, construction, expansion, and reproduction of beliefs in societies, whether religious with mythical points of reference or secular with rationalistic, historical requirements.[1]

Mohammed Arkoun

There was a time, now almost a generation ago, when there were relatively few ethnographic data collected by trained anthropologists working in Islamic societies, when anthropological approaches to religion focused on "primitive" non-worldwide religious systems, and when those who did devote their scholarship to Islam did so almost exclusively as exegetes of texts. This was the academic setting that *Islam Observed* addressed, Ernest Gellner embellished, Fatima Mernissi skirted, and Akbar Ahmed derided. Edward Said, I should add, covered it and condemned it. Fortunately, there is now a sizeable presence of ethnographic analysis of Muslim societies; unfortunately, it is rarely known or cited outside of the narrow confines of specialized subfields in anthropology. Islam, mainly one geographical zone of it, appears in several summations of Middle East anthropology, but no one has yet charted the intellectual trajectory of an anthropology of Islam as such.

Perhaps there is no need. Imagine the absurdity of writing an anthropology of Christianity by tracing all the ethnography conducted in "Christian" contexts. What would such a far-fetched and novel text be called? *Christianity Observed, Christian Society, Beyond the Bread and Wine, Discovering Christianity*![2] What precisely would an anthropologist contribute to a topic that has several fields devoted exclusively to it? That Christianity in Spain is carnivalesque and its variants in Lebanon are confessionally mired? That flux and reflux made medieval serfs into Victorian Sunday school teachers? That *the* Christian view of woman is *vagina dentata*, not the Holy Virgin Mother? That a Christian could write about his own dogmas with an empirically intact conscience? I suggest that simple essentializing of the long history of the faith into ideal types, beyond repeating the obvious sectarian splits, offers nothing new. It is easy to create unity out of diversity but seldom does it serve an analytical purpose.

It is my main contention that the selected examples of now "classic" anthropological approaches to Islam obscure the effective understanding of Islam as a cross-cultural faith embedded, quite deeply at times, in numerous cultural traditions. My emphasis throughout has been on the negative, highlighting what has gone wrong textually. To carry the metaphor further afield, we have just walked through—far too briskly, at times—a minefield of interpretive problematics and flawed rhetoric. My ultimate aim is not to die a martyr's death on an academically inclined and intellectually predefined battlefield, but to properly detonate the unexploded myths and seductive biases. Iconoclastic deconstruction will never get an author to the right side just because it avoids the wrong side. Now comes the really difficult part: charting a course of safe passage that will stay clear of the same, and perhaps irritatingly resistant, fallacies so prevalent in the texts of Geertz, Gellner, Mernissi, and Ahmed. Having objected to the obscuring of Islam, what is it that anthropologists have, can, and should do to improve their perspectives and methods for a more enlightened but less enrooted understanding of the religion of Muslims?

I am in the enviable position of not having to invent new wheels, but rather to spin some of the main points that a growing number of colleagues in the discipline have already made or at least have strongly hinted. I cannot claim to have surveyed a comprehensive range of these sources, let alone those appearing in languages other than English; nor do I wish to overburden the narrative with an annotated listing of what has been done and who said what. I prefer to shift from disputation by continuous narrative flow to a set of targeted responses to a few key questions. This stems in part from a personal prejudice that posing the right questions should be as challenging as proposing answers.[3] I am also enamored of the contemporary Muslim philosopher Mohammed Arkoun's pragmatic approach in *Rethinking*

Islam, a primer in question/answer format on what many Western commentators think Muslims incapable of doing. The questions I frame here address the anthropology of Islam; they are intended for anyone interested in what anthropologists can do, as well as a call to action for colleagues in my discipline. It is, after all, an informed inter-disciplinary appraisal, more than any single academic field of view, that holds the most promise.

What is the Difference Between the Anthropology of Islam and the Sociology of Islam?

It is the character of lived experience I want to explore, not the nature of man.[4]

Michael Jackson

Whether or not an approach to religion is anthropological or socio-logical is a bit of a red herring. To a certain extent the answer is as trite as the discipline in which a researcher has been trained. But the interchange of labels is too rampant to be dismissed as simple cross-border interchange. Consider that French scholar Jean-Pierre Digard provides "perspectives anthropologiques" in a French journal of "sociologie," while calling what he does "ethnologie."[5] In France Jacques Berque and Pierre Bourdieu teach "sociologie." In Britain a number of social anthropologists regard what they do as a type of sociology, a notable example being Ernest Gellner.[6] Since Gellner was trained as a philosopher and harbored lifelong suspicions of any notion that could be called Wittgensteinian, I suppose one category is as good for him as another. American academics are generally more disciplined. Clifford Geertz is an unabashed anthropologist, although he relies to a great extent on sociologists like Weber and Parsons. Even in formal ethnographies, the bread and butter of anthropologi-cal communication, the distinction can be fuzzy. Dale Eickelman, who conducted an ethnographic study of a pilgrimage center in Boujad, Morocco, identified his primary goal as making "sociological sense."[7] Given that most readers have a relatively clear idea of what sociology is about but little knowledge of what anthropologists do, the word choice may in fact be pragmatic rather than programmatic.

Introductory textbooks teach either anthropology or sociology, even though professors of each often draw on both disciplines in their assigned readings. In crude terms, anthropologists are usually associ-ated with doing research in exotic venues and sociologists are more likely to hang around institutions and hand out complicated surveys that require a knowledge of statistics to decipher. Robert Hefner, for example, notes that "Islam" was the focus of sociologists in Indonesia before the late 1970s; anthropologists dealt with the peripherals.[8] Thus, an anthropologist must do ethnography by living in an Islamic context, while a sociologist could just find the university library and play with data sets. Neither type of social scientist, it should be stressed, habitually spends times with indigenous texts, apart from an over-appreciation of the "sociological" demeanor of someone like Ibn Khaldun.

Who has the advantage? A specialist in Religious Studies or History would say neither. The choice of theory and method—a very wide selection indeed at the start of the third millennium—depends to a large extent on the questions being asked. Let us say the subject is the education of Muslims in religious schools in Pakistan. An anthropologist would likely apply for funding to go to a Pakistani village or town to observe behavior in schools, talk to the people involved on a daily basis and conduct informal interviews. A sociologist might gravitate more to the demographic statistics gathered by the government on pupils and teachers, create a survey questionnaire and analyze the institutional structure.[9] Meanwhile, the Religious Historian would rightly inquire, "But can you actually read in the original what the students are reading and comparatively trace the content through the textual tradition?" The issue should not be who would achieve a "better" understanding, given the obvious limitations in specialized training, but how the available tools could be applied and adapted by an individual scholar.

The terminological confusion is compounded by that fact that a nontheological, or at least nonexegetical, study of religion is commonly labeled "sociology," even by anthropologists. I suspect this is due in large part to the sociological credentials of scholars like Durkheim and Weber, who are as likely to be read by anthropologists as by someone trained specifically in the modern discipline of sociology. Marx himself should be included in this intellectual trajectory, although reducing his meager contributions to the study of religion as "sociological" seems as uneconomical as it does Whiggishly self-serving. As Talal Asad has warned, there is a danger in applying indiscriminately to Islam such widely distributed concepts as Durkheim's sacred and profane or Weber's ideal types.[10] The issue is not how useful such concepts could be, but the need to recognize the cultural specificity of the contexts in which they are commonly made. There is so much debate about the methodological problems in past sociological models of religion that borrowing contested terms may simply beg the theoretical questions. Arguing about religion, it might be said, readily becomes the opiate of social scientists, whatever their formal training.

What do Anthropologists do Differently when They Study "Islam"?

An encounter with lived Islam challenges conventional definitions of religion as consisting primarily of beliefs and practices set apart from everyday life.[11]

Carol Delaney

Archaeologists unearth material relics, epigraphers decipher inscriptions and historians read manuscripts. The primary source material for an ethnographer is what people do and say in the ethnographer's presence. One need not be trained as an anthropologist to observe Muslims or describe their behavior. There is a large and potentially useful corpus of description left in print and archives by travelers. Victorian Edward Lane, for example, spent considerable time in Egypt during the first half of the nineteenth century and eventually wrote *An Account of the Manners and Customs of the Modern Egyptians*, a descriptive account of just about everything a curious English gentleman abroad might find interesting. Much of this is valuable documentation, but it is no more "ethnography" in the modern sense than Darwin's *On the Origin of Species* substitutes for modern genetics. If the primary field method of anthropology is reducible to mere observation, it is open to anyone willing to travel and reflect. Clearly, there should be more at stake. Just as critical historiography involves more than merely reading a historical text, so ethnography goes beyond writing down observations of curious customs.

Scholars outside anthropology often elide this distinction. "Much of the early ethnographic work on actual Muslim communities was done by Christian missionaries," writes the distinguished historian of religion Richard Martin.[12] The missionaries, by and large, were motivated to convert the Muslims they observed. If you are interested in what part of the sycamore tree Palestinian peasants found edible, then consult a missionary ethnography such as *The Land and the Book* by Rev. William H. Thomson, originally published around the same time as Lane's *Modern Egyptians*. But be prepared for sentiments like the following:

> Change the state of society (and in many places it is being changed), educate the females (and the males too), let the community be pure from Moslem and heathen mixtures, and trained to free and becoming social intercourse, and then neither men nor women will think of veils and screens, nor need these apostolic directions in their exact letter.[13]

Being an ethnographer does not make a scholar objective, nor erase cultural presuppositions, but it is certainly a leg up over being a blatant religious partisan or armchair theorist.

A proper case in point is Daniel Bradburd's *Being There: The Necessity of Fieldwork*, an autobiographical account of living with the Komachi nomads of southeastern Iran in 1974 and 1975. Like many anthropologists who have done research in the Middle East, Bradburd did not set out to study Islam. His academic focus was set on household economy among pastoralists, but the Komachi happened to be Muslims as well as nomads. The older travel literature describing the Komachi spoke of them as fanatical Persians who hated Christians, while an earlier ethnographic study of nomads in the general area made it seem as though they would be irreligious. Yet Dan Bradburd and his wife, Ann Sheedy, found that the Komachi "truly lived in a Shi'ite world," not only in the outmoded "little tradition" sense, but with direct ties to formal religious practice.[14] Bradburd's account, readably anecdotal, is a passionate defense of participant observation as the anthropological method *par excellence*. Being there was not an end in itself, not an act of domination—certainly not from the standpoint of the Komachi—but an opportunity to build a "meaningful model" of what they saw from within a give-and-take situation where that model could be mulled over and corrected. Imagine a historian who could go back in time and interview an author, even witness the events being described in a text: that is the potential of ethnographic fieldwork for understanding how Islam is lived in a specific social setting.

Perhaps the most original contribution an ethnographic approach can offer is charting how beliefs and ideas are put into practice: not how they are supposed to be or should be, but how they unfold in an observable manner in one small place at one particular time. Ladislav Holy, for example, studied the Berti of Northern Darfur in western Sudan for a little over three years between 1961 and 1986. As an African society far off from the metropolitan centers associated with "great tradition" or normative Islam, the Berti could all too easily be dismissed as idiosyncratic syncretists, mixing Islam with earlier indigenous religious beliefs. Holy found that even the most pious members of the group were tolerant of actions that Islamic precepts would appear to prohibit. As an example, most Berti did not think the Quranic prohibition of wine applied to the local millet beer. While local religious scholars abstained from drinking beer themselves, they never preached against it and would even offer a bowl to guests. "The tolerant attitude to the sinful ways of others derives from the fact that the pious do not feel themselves in any way implicated by the acts of others," explains Holy.[15] Yet, other local customs would be railed against by the pious, especially when these hindered their ability to perform religious duties. Only by being there, observing behavior and its consequences, could the anthropologist begin to unravel the local meaning of behavior responsible to textual precedent.

Holy, like many recent ethnographers, follows the Malinowskian dictum to map out what is happening from the native point of view. Rather than dismiss Berti practice as "little" in contrast to a "great" tradition that hardly exists in the local context, his focus is on the various ways in which Islam is lived in concrete, pragmatic terms. This requires building up from observation of specific behavior and following existing debates rather than measuring local practice according to a textualized ideal of what should be the case. "For the student of religion who does not want to change, condemn or justify existing beliefs and practices but to understand what they mean to those to whom they belong," argues Holy, "there is little point in classifying them according to how closely or remotely they approximate the ideal."[16] Muslims, like everyone else with human credentials, disagree in practice. The point is not whether they should, but how the artificial category of "religion"—certainly a foreign notion to the Berti themselves—masks the complex negotiation of individual and communal concerns on a day by day basis.

A major objection raised about analyses of local versions of Islam is that there appears to be an endless number of versions. Scholars who want to see the broad picture complain that ethnographic snapshots often confuse the issue. Certainly it is problematic to use the study of a single community as panoptic for a region, ethnic grouping, or even Islam as such. This is a major criticism of the essentializing of Geertz, Gellner, Mernissi, and Ahmed. But, sampling problems and overstretched interpretations aside, anthropological studies are capable of bringing new and original information to the ongoing debates. As John Bowen cogently argues, "local studies give us a window onto the rhetorics and forms that mediate between 'local' and 'translocal' phenomena."[17] I do not think there is a danger of knowing too much about variations in Islamic practice. The critical issue is balance, using observable social contexts to rein in the tendency to substitute the ideal for the real.

Ethnography is not a panacea for essentializing, but it does offer an important corrective at times. It is both process, an interactive method of getting information, and a product that literally re-presents that information in a frame not of the native's choosing. In the process the modern ethnographer is a cross between photographer and artist. A photographer usually attempts to capture a scene as it exists at a given moment, to freeze reality in a frame for remembrance. But the artist reinterprets the observable reality in order to highlight some aspects and ignore others, to imagine rather than duplicate an existing image. Photographing social behavior may be objective in intent, but the interpretive lens of the ethnographer always filters what seems so natural at a given moment. In a sense, participant observation tends to be objective only as the observer becomes consciously aware of the cultural and discipline-training biases in his field of vision. Lived experience cannot be simply observed, since only certain events will be recorded and judged to be relevant. There is no one program for deciding which observations to draw on, certainly not What Always Works 101 in graduate training.

Consider one specific and mundane event from my own fieldwork experience in rural Yemen. One local man did not know, let alone comprehend, the exact words in the prayer ritual, despite the fact he was a native speaker and had been brought up as a Muslim. When he prayed, it was obvious to the people around and to the ethnographer that he would mumble over passages that were easy to memorize. How then, should I as an outside observer of behavior represent this? From what I could see and learn through conversation, and that was hardly an omniscient perch, his fellows were tolerant of this behavior, did not attribute it to unbelief and accepted the fact that he was sincere even if not very bright. In Redfield's view this would be a prime example of little tradition mechanics, a case of not being literate in what the great tradition establishes as normative. For a Muslim not from the area the man's ignorance might even be taken as a sign of heresy to be corrected.

How should this discrete event involving a particular individual be represented? Unlike my sociological colleagues, I have no access to statistical data on how many men in the local sampling area were ignorant of the words as they prayed. Not being a psychologist, I cannot say what traumatic childhood event might have been blocking his memory or informing his attitude? Since I am not an *imam*, it does not offend me morally that a man could sincerely follow the ritual and do so improperly. I could, as a good historian, examine the available *fatwa* collections and see if such a case has a precedent in Islamic legal tradition; this would at least make an impressive footnote. As an ethnographic observation it may in fact be so irrelevant as not to warrant inclusion in a discussion of local ritual behavior or it may be a major bit of evidence in arguing for how individual variation plays out in ritual behavior. Whether to represent this event and how to do it are clearly subjective choices. The datum does not speak for itself, even if I allow it to become a published datum for others to interpret for themselves.[18] Once presented in print, even as I do so here, I can hardly anticipate all the possible ways in which my representation will itself be represented. Old stories about parrots, as the infamous Bororo case well illustrates, fly even in the face of reason.

What are the Major Challenges to Successful Ethnographic Fieldwork?

To me the field experience is far more like a sharp blow to the head or a large spoonful of horseradish; it is a process that marks those who have been through it.[19]

Daniel Bradburd

Anthropologists observe Muslims. Ethnographic fieldwork is done at tree level, no matter the view of the forest an individual researcher brings with him or her. This requires the ethnographer to know how to distinguish one tree from another. Essential to this process is the ability to communicate with people in their own language. Working through an interpreter can yield bits of information, but the extra layer of filtering removes the creative dynamics of communication. Most anthropologists enter the field with a minimum of language training; nor are all ethnographers competent linguists. The four books examined in this study show the range of possibilities. Geertz learned enough Indonesian to function in Java and Bali, but there is no indication he developed more than a smattering of colloquial Arabic; none of his sources in *Islam Observed* are direct Arabic or Berber sources. In *Saints of the Atlas*, Gellner makes no mention of how be obtained his information from Berber informants, apart from the admission that he possesses "a bad ear and no linguistic training."[20] His bibliography suggests that the majority of sources consulted were in French. Fatima Mernissi, a Moroccan Arabic speaker, and Akbar Ahmed, a native of Pakistan, clearly had an advantage in their fieldwork by collecting information in their respective native languages. I do not suggest that being a native speaker makes one a better ethnographer, but the potential to appreciate nuance is theoretically greater if one actually knows the language well.

The student of culture in the field often faces a problem not to be found in a university library or communicating with peers: culture shock. Until the last quarter of the twentieth century, feelings of inadequacy, disappointment, and outright depression were sentiments rarely discussed by ethnographers. Apart from Colin Turnbull's overriding contempt in *The Mountain People* for an African tribe that failed to measure up to his beloved *The Forest People*, bad fieldwork experiences are easier to ignore than admit. Gellner's *Saints of the Atlas* reads like a holiday write-up, which in a way it was for the English professor on Easter, Summer, or Christmas vacation.[21] Geertz romped through the private diaries of Malinowski, but saved his own autobiographical excursus, carefully crafted for publication, until quite recently. His student, Paul Rabinow, who ended up in

Morocco by default, reflected on why fieldwork was not what he expected it to be. Between the stoic silence of the old guard and the whining of the young and unprepared, there are few published ethnographies on Muslim societies that address the natural state of shock in leaving the familiarity of home for the exotic unknown of a fieldsite.[22]

Culture shock is not a disease; nor should it be seen as a failure to adapt. Part of the process is physical in a way beyond conscious control. After only a few weeks into my intitial fieldwork in Yemen, in 1978, the local "flora" invaded my intestines with a vengeance that left me on my back several weeks. I thought I could just let nature take its course, but my wife wisely dragged me to a local hospital. The Yemeni patients who had been waiting all morning took one look and immediately waved me to the front of the line for immediate attention. Graduate training in anthropology rarely prepares for the inevitable health battles. Old habits of decorum die hard, as Daniel Bradburd can attest. In the middle of the desert he awoke one night with a bad case of diarrhea. He ran for the outhouse, a gully some fifty yards away, but his bowels only held out for the first thirty. When he tried to explain to his wife what had happened, she shook her head: "You were running for the gully," she echoed. "It was three o'clock in the morning, you had five hundred thousand square miles of desert to take a shit in, and you shit in your pants because you had to go in the gully. What makes the gully different from the rest of the world?"[23] Consider that some of us arrived in the field never having been through a day without a nearby roll of toilet paper; even my father's depression-era outhouse had a Sears catalogue handy.

Anthropologists do not as a rule go native, but the ideal goal is to experience life as the locals do. Gellner noted that anthropologists are at home in villages, but it is not clear for how much of his fieldwork time villages really were his home. Whatever anthropologists say after the fact, settling into a local fieldsite is rarely an easy task. When my wife, Najwa, and I chose the location for our field site in highland Yemen, we asked a local sayyid notable for help in finding a house, or at least a room, in one of the local villages. This man lived in a relatively new and grand house with piped water from a spring, flush toilets and a private electricity generator. He immediately offered us a room in his home with a separate entrance, piped water from a spring, a flush toilet, and occasional electricity. But we had come to live in a village, so we politely demurred and dutifully followed him on a walking tour of the nearest hamlet. Despite the observable fact that there really were no habitable dwellings for rent and that Najwa, who was also an anthropologist, would have to walk up and down narrow stone steps on a steep cliff several times a day to fetch water in a plastic pail—balanced delicately on her head—we diligently tried to find a home in a village. Pragmatism soon won out; we moved in with the sayyid, thankful for the delight of pseudo-bourgeois amenities. This was probably one of our more sensible decisions during the entire fieldwork process.

Because fieldwork is a highly personal experience, there is no magical formula for success. It helps if the local people are willing to put up with a stranger in their midst. It does not help to walk into a village and assume that everyone is a potential informant with nothing better to do than sit and answer questions, speaking very slowly in the most basic elementary Arabic. Living in a remote village can be lonely; being the constant center of attention for inquisitive onlookers and myriad curious children can wear down even the most resilient demeanor. As an American, Dan Bradburd was nurtured on the inalienable right to be alone at times. Yet among the Komachi nomads he discovered that such a seemingly natural claim for private time and space struck his hosts as a problem and in extreme cases could cause a rupture in the social fabric. For the Komachi "aloneness was a thing to prevent rather than to promote."[24] As Bill Young felt, while trying to write up notes in a Rashaayda tent, "Why should I prefer to sit and write when I could talk to human beings instead?"[25] Doing fieldwork required a certain degree of undoing the givens in his own worldview.

Adjusting to intrusions of cultural space is often not half the problem of dealing with the emotional consequences of joining a community where death and illness take a prominent role. One of the most painful memories of my own experience in Yemen was returning one day about two years after my fieldwork and spotting a good friend. On a previous visit I had taken a photograph of him with his young daughter. It was such a striking picture that I had it enlarged while back in the states. Normally, I followed the routine of inquiring about family and friends, but this time in my own excitement I rushed to my friend and handed him the picture. As he glanced at the image, I knew immediately from the pain in his eyes that something was wrong. I later learned that his daughter had died, run over in a tragic car accident in the village, only a few weeks before. Najwa and I often gave photographs we took of village people back to them as gifts, which we would sometimes see on a wall as we visited homes. We eventually noticed that when a person died the photograph was almost always taken down or turned to face the wall. It was only after this experience with my friend that I understood the reason. In a society where the dead are not embalmed and displayed for mourners, pictures of the recently deceased invoke emotional pain. The quotidian world bedevils the best of intentions.

Is there One "Islam" or Many "islams"?

Anthropologically, the problem now is to find a means of understanding that order which reaches the desired level of universality without diluting or destroying the significance of this diversity and the richness of meaning in human experience.[26]

Abdul Hamid el-Zein

"Islam" as such teaches us nothing.[27]
Omid Safi

The essential problem in the study of Islam is precisely that: essentialist reduction of a diverse religious tradition across cultures into an ideal essence. To clarify my iconoclastic leanings, let me endorse the caveat given by Talal Asad: "The argument here is not against the attempt to generalize about Islam, but against the manner in which that generalization is undertaken."[28] I do not wish to be dismissed, even nominally, as a nominalist. My problem is with "Islam" with a capital "I." Geertz thinks he has observed it, Gellner has theorized it into a philosophical whole, Mernissi attacks what it does to the Muslim female, and Ahmed discovers it all over again for his English readers. In a provocative article published a quarter of a century ago, Muslim anthropologist Abdul Hamid el-Zein wondered in print "if a single true Islam exists at all."[29] Unlike Akbar Ahmed, this was not an attempt to Islamicize anthropology or probe the theological ground foretold by Evans-Pritchard, but a challenge to scholars who blithely assume the existential "truth" of concepts. "But what if . . ." asked el-Zein, analysis of Islam "were to begin from the assumption that 'Islam,' 'economy,' 'history,' 'religion,' and so on do not exist as things or entities with meaning inherent in them, but rather as articulations of structural relations, and are the outcome of these relations and not simply a set of positive terms from which we start our studies?"[30] If so, he reasoned, it would do no good to start with a textbook version of the five pillars, Ibn Khaldun, David Hume or Max Weber, because all this is what Islam is supposed to be. For el-Zein, true to his anthropological roots, it was important to start with the "native's model of Islam" as it is articulated in a given social context. This is not because the native is "right," a nonsensical term for non-theologian el-Zein, but in order to see how Muslims adapt what analysts call "religion" to everyday life.

It is worth revisiting el-Zein's argument, not only because it tends to be ignored or misunderstood, but as an important reminder of what it means to study Islam ethnographically.[31] Among those who miss the point and stumble on this easily decontextualized phrasing is

Talal Asad, who begins his brief lecture on "The Idea of an Anthropology of Islam" by dismissing el-Zein's "brave effort" as "unhelpful."[32] Asad asserts that el-Zein is the victim of a logical paradox: claiming that diverse islams are equally real and at the same time that they "are all ultimately expressions of an underlying unconscious logic." Sensing the taint of "Levi-Straussian universalism," Asad faults el-Zein for dissolving the very analytical category, Islam, that he is searching for. I think there is less disagreement between the basic arguments of el-Zein and Asad than this entails. El-Zein would have agreed heartily with Asad's bottom line: "It is too often forgotten that 'the world of Islam' is a concept for organizing historical narratives, not the name for a self-contained collective agent."[33] The real problem, unstated by Asad, is a disagreement over the nature of "culture," arguably the most debated and fought over concept in the history of anthropological theory.

Following Asad's criticism, Fadwa El Guindi likewise rejects the idea of one versus many islams, while Carol Delaney thinks such a model denies any meaningful function to Islam.[34] Robert Launay suggests that el-Zein's insistence on multiple islams is "in essence, to make a theological claim," but this assumes el-Zein, a Muslim, is denying the validity of his own faith.[35] The point is not that there can be no Islam, but that such a concept serves little anthropological purpose; this hardly turns the anthropologist into a theologian by default. As Joel Robbins observes, the arguments used against el-Zein's concept of multiple islams are not "particularly sophisticated in theoretical terms."[36] I suggest that critics read into el-Zein's argument more than he was actually saying, certainly far more than he intended.

Advocating a "phenomenological" approach at the time, el-Zein believed that underlying the diverse "contents" of cultures was an embedded "logic" in the very nature of culture. Thus, there is a sense in which both the anthropologist and the native, although from different cultures content-wise, share "a logic which is beyond their conscious control."[37] Unfortunately, el-Zein did not elaborate in this brief review of several texts about Islam what this logic entails; he passed away soon after the article was published. The critics, however, ignore el-Zein's practical application of this theoretical frame in his excellent ethnography on Lamu. The logic he was talking about refers to the structured relation of symbols in the narratives and speech of Muslims he observed and queried. Religious symbols, like Muhammad, Adam and Eve, or the Quran, are not approached as "entities nor fixed essences" but rather serve as "vehicles for the expression and articulation of changing values in varying contexts."[38]

In the context of Lamu, for example, he analyzes the ways in which masters and slaves, decidedly different social categories, appropriate the symbol of the Prophet Muhammad as light (*nur*) to articulate opposing worldviews. Influenced, but not blindly so, by the structuralism of Claude Lévi-Strauss and interpretive anthropology of Clifford Geertz, el-Zein was seeking a way to go beyond the surface functions to a deep structure of the religious ideology.

Theoretical precision here is somewhat of a moot point, for el-Zein's main argument is that "Islam as an expression of this logic can exist only as a facet within a fluid yet coherent system; it cannot be viewed as an available entity for cultural systems to select and put to various uses."[39] Of course it had been viewed this way by scholars of multiple disciplines, which is one of the reasons he was critical of previous studies. El-Zein was indeed arguing that the notion of Islam "without referring it to the facets of a system of which it is part, does not exist." The notion did exist in the minds of many writers and certainly among theologians, but as an essentialized and pregiven definition it would be "extremely limited in anthropological analysis." Thus, when he states at the conclusion of his review that Islam does not exist as "a fixed and autonomous form referring to positive content which can be reduced to universal and unchanging characteristics," it is hard to imagine why any scholar other than an apologist for Islam or ardent antagonist of Islam would find cause to disagree. Consider the recent call of Omid Safi—"For better or worse, in truth or ignorance, in beauty and hideousness, we call for an engagement with real live human beings who mark themselves as Muslims, not an idealized notion of Islam that can be talked about apart from engagement with those real live human beings."[40] Nor is el-Zein confining his theoretical approach to "Islam" alone, but to the very notion of "religion" as a meaningful category in and of itself.

Ethnographic research revolves around localized "islams," as el-Zein would put it, diverse cultural contexts in which Muslims live out something both the natives and the anthropologist invariably refer to as "Islam." The Yemeni tribesmen I lived with did not conceive of their faith as one of the numerous ways in which Islam was practiced; as far as they were concerned they were practicing true Islam or at least something very close to it. This is no different from the fundamentalist Baptist community I knew as a child; the church members believed they were practicing true Christianity, not just being members of one mid-twentieth-century sect among many. At ground level there is a sense that the religion practiced is the right one or that it can be corrected to be so. Specific practices might change, beliefs may be dropped or added, but faith at the bottom is only meaningful if it can be lived meaningfully as more than a local phenomenon. Thus what the anthropologist defines as just another islam is invariably seen by the practitioner as an attempt to do Islam. The issue is not whether this Islam exists; if there were no concept there would be no meaningful distinction to being Muslim. Theologians have no trouble with an idealized Islam, but should ethnographers among Muslims operate with this conceptualized Islam as a given, as something meaningful in itself, apart from its local appropriation?

The importance of el-Zein's reorientation of anthropological concern with Islam stands out against the analytical tendency in the four seminal texts examined earlier to explain the Islam behind all the local versions as a master blueprint. Both non-Muslims, especially those carrying apologetic or religiocentric baggage from their own cultures, and Muslims have had a stake in defining a single, "true" Islam. More recently, with the mainstreaming of political correctness, this trend has reversed and it is almost an obligatory starting point in scholarly treatment to acknowledge that "There is no monolithic Islam."[41] This important recognition does not stop most textbook treatments from summing up Islam as a sacred text, prophet, and set of pillars. Moreover, there is a tendency to treat Islam as visibly practiced in the Middle East to be *the* Islam, as though the fact that more than three quarters of Muslims live outside the region is not relevant. This is seemingly inevitable when pedagogy demands generalization and seduces scholars to overlook nuance.

Similarly, those who view Islam—however misunderstood—as a threat, need it to be a homogenous target, a straw religion easily denounced and demonized. It is not hard to find such a fallacy fetish among rightwing Christians, who go so far as to equate Islam as a conspiracy of Satanic dimensions. Shortly before the first Gulf War, fundamentalist cartoon evangelist Jack Chick published a serious comic that uncovered Islam as a religion originally created by the Catholic popes to gain control over the Arabs. "Satan was determined to block the gospel of Jesus Christ from reaching the children of Ishmael. By using the Vatican, Satan closed the door for centuries, depriving the Arabs from hearing about the Light of the World," confides an ex-Jesuit, who is said to have read all about it in the depths of the secret Vatican archives.[42] The unwary reader is told that Muhammad was a stooge seduced by a former nun, Allah a moon god, Islam was spread by a blood-soaked sword and—most significantly—that Muslims are taught to only see bible-believing missionaries as devils.

While bigots are rarely convinced by rational arguments, serious scholarship should try to evolve beyond the stasis of prejudice. A really pathetic fundamentalist anti-Islamic diatribe like Robert Morey's *The Islamic Invasion* is logically trashed by a readable scholarly text such as John Esposito's *The Islamic Threat: Myth or Reality?*[43] But being an established scholar does not mean that a mutually acceptable objectivity is always reached. Consider the post-9/11 article by Bernard Lewis in *The New Yorker* on "The Revolt of Islam."[44] While not overtly condemning Islam, a reader might reasonably conclude from the information in the article that Islam is not only in revolt but is in many unpleasant ways rather revolting at the present time. "The Muslim peoples," states the historian, "like everyone else in the world, are shaped by their history, but, unlike some others, they are keenly aware of it."[45] This is the reason, Lewis assures us, that the anti-American war talk of Osama Bin Laden and the actions of Muslim terrorists resonate in the world of Islam. But Lewis misses the point here by assuming Americans only treat "history" as water under the bridge, especially the violent history of religious wars among Christians that plagued Europe for centuries. Bin Laden may cave-dream of a return to seventh-century Muslim unity, but the "history" that impels his rhetoric is surely recent placement of American troops in Saudi Arabia, America's political seduction of self-serving Arab leaders, and continued United States support for Israel's oppression of Palestinians. Not given to citing postmodern critique of the very establishment he represents, Lewis assumes the Muslim world has not even managed to embrace modernity. Yet, Bin Laden was not born a nomad and is unimaginable without stinger missles and video sermons.

Islam is not in revolt except as a foil for those who prefer to look at individual acts as mere pieces of a Leviathanesque essence. Palestinians revolt, desperately coating an Islamic veneer over political acts for confessional comfort. Taliban revolt, as anthropologist David Edwards documents in his historical reconstruction of Afghanistan history through ethnographic interviews.[46] Individual Muslim women revolt by taking off the veil, or in some cases, by actually putting on the veil. What ethnography can offer is precisely what abstract pronouncements like the West against the East or us versus them scenarios of Bernard Lewis, or Edward Said, are lacking. Thus, the buildup of ethnographic knowledge about how Islam is currently practiced in widely varying contexts can only be for the good in getting beyond the ongoing politics of blame.

How Should Anthropologists Study the Great Tradition of Theology?

The fact is that proper ethnographic characterizations of local Islam require familiarity with Islam's textual and normative sources.[47]

Robert Hefner

The theoretical question 'What is Islam?' and the theological question 'What is Islam?' are not the same and should not be conflated.[48]

Ron Lukens-Bull

When Abdul Hamid el-Zein conducted his search for the anthropology of Islam, he wanted to do it "beyond ideology and theology." For el-Zein, but definitely not for Akbar Ahmed, the anthropological approach to Islam was diametrically opposed to the theological. This was far from a confession of disbelief, but a recognition that theologians—whether Muslim or not—have "different assumptions concerning the nature of Man, God, and the World, use different languages of analysis, and produce different descriptions of religious life."[49] Ironically, as el-Zein noted, both theologians and anthropologists often end up by validating boxed versions of an essentialized Islam: theologians condemn local variants from the orthodox norm they validate, while anthropologists tend to reduce local practice to superstition and syncretism that distort the assumed pure essence of the religion. The anthropologist is not likely to study Islamic theology in order to determine its spiritual truth, but it is almost nonsensical that an ethnographer would attempt to study Muslims without knowing seminal texts like the Quran, *hadith* collections and relevant legal texts.[50]

Talal Asad suggests that theology, at least as Muslims define it for themselves, cannot be ignored in anthropological analysis. "If one wants to write an anthropology of Islam," he suggests, "one should begin, as Muslims do, from the concept of a discursive tradition that includes and relates itself to the founding texts of the Qur'an and the Hadith."[51] Asad prefers to define Islam as a "tradition," a fairly open-ended concept within and without anthropological writing. But it has a more closed nuance for Asad, even though he does not probe his own theoretical influences. "An Islamic discursive tradition," he explains, "is simply a tradition of Muslim discourse that addresses itself to conceptions of the Islamic past and future, with reference to a particular Islamic practice in the present." This is anything but simple, even though his main point is well taken: Islam cannot be properly understood in a synchronic mode. But the potential corpus of Muslim discourse is vast and varied. Interpretation of the Quran as it is applied to daily life has been widely contested, as has the validation of which statements by the prophet can be considered authentic. How then, at least in a single lifetime, is an ethnographer able to make sense both of the behavior being observed and the tradition being invented against the overall discursive potential of formal texts and oral narrative? Can the native view ever be adequately understood without becoming, in a practical rather than a practicing sense, a native oneself?

The majority of anthropologists who have worked with Muslims probably have not read the entire Quran, certainly not in Arabic. Few would have the training, let alone the desire, to do so. Yet, it is not unusual to find illiterate Muslims who have memorized large portions of the Quran or are, in a sense, walking oral texts of their faith. When Robert Redfield long ago advocated a new approach in which anthropologists observing the little tradition would work cooperatively with historians who studied the great tradition, he was making an important plea for interdisciplinary research. Team research sometimes works, but it can also degenerate into knowledge manipulation by committee; the King James Version of the bible is a relevant example. Theoretically, there is no reason why an ethnographer cannot at the same time be a competent linguist and read the same texts known to the people he lives with. The anthropologist need not be a theologian, trained in exegetical method, but an awareness of the textual sources available for Muslims is clearly an advantage, at times a necessity, for analyzing how an eminently textualized faith plays out in the local community. Nadia Abu-Zahra, Richard Antoun, Sayed el-Aswad, and Abdul Hamid el-Zein, all native Arabic speakers, illustrate this potential in their ethnographic analyses of sermons and storytelling.[52] I do not suggest that one has to be a native speaker, although clearly it is an advantage, since there are anthropologists who achieve a competent knowledge of the relevant language and corpus. Robert Launay's study of Dyula sermons in the Ivory Coast and Brinkley Messick's contextualization of Yemeni legal texts illustrate this well.[53]

The danger is not in knowing too much about normative Islam, an absurd false dilemma, but in how to apply what is said in texts to the contexts of use or allusion. The Quran is a case in point. Like all sacred scriptures that have been revered over centuries, there are many possible interpretations of Quranic meaning. No verse speaks for itself, despite what might appear to be a clear statement to an outsider. Proper understanding is complicated in the case of Islam, because Muslims are taught that the Quran must be read in its original Arabic. A translation is thus always an interpretation and as such not equatable with the original; there is no King-James-Version-only mentality among non-Arab Muslims. This is not to say that translations are unimportant in influencing how individual Muslims think. Among the Sama of the Philippines, Patricia Horvatich found that local people insisted on having the Quran in English or their local language so they could better understand it.[54] This was even encouraged by local Malay missionaries. Translations of the Quran abound, not so much as vehicles for God to convict the sinner as in Christian missionization but as helps for the believer to eventually master the Arabic needed for true understanding. Beyond theology, which may be on a superficial level among many Muslims, the Quran is of paramount importance as a paradigm for phrasing, even in translation. Carol Delaney demonstrates that a Turkish gender model of male and female as "seed and field" relates to the symbolism of a Quranic passage.[55] Nadia Abu-Zahra notes that the local Tunisian use of *rahma* (literally "divine mercy") parallels the metaphoric usage in the Quran for rain.[56] The men and women I talked with in Yemen routinely sprinkled everyday speech with verses from the Quran or relevant traditions alongside culturally specific proverbs and colloquialisms.

Properly noting the relevance of the Quran and other textual sources of Islamic tradition requires more than simple comparison of context to text. As Fadwa El Guindi laments in her analysis of how "veiling" has been misrepresented, "A few Qur'anic *Suras*, particularly the ones pertaining to the subject of women, are routinely and uncritically referenced from secondary sources, their English translation unchallenged, and their meaning presumed."[57] This accounts for the bizarre Freudianization of the Moroccan ritual of sacrifice by Combs-Schilling referred to earlier, or Mernissi's assumption that a medieval scholar's prurient prose about sex should be deemed paradigmatic of a homogenized Muslim gender ideology. Just as ethnography in the modern sense involves more than simply being in some "there" and writing about it, so analysis of religious and legal texts goes beyond knowing how to read *per se*. Anthropologists should read Islamic texts, but with a willingness to learn rather than an eagerness to find some text proof for a point they want to make.

Ethnographic fieldworkers in Muslim societies paid little attention to what Redfield called the "great tradition" until the last two decades of the twentieth century. As a result the idea that some Muslim groups, especially nomads, were observed to have a "paucity of ritual" became a kind of anthropological truism in which something other than religion could serve the function often attributed to religious ritual. In criticizing the unwillingness of several of his colleagues to recognize rituals that Bedouins themselves would define as religious, Emrys Peters argues that just because some Muslim groups are not as "piously punctilious" as others does not mean they are not connected, perhaps even more meaningfully than "any number of genuflections" with the core of Islamic beliefs.[58] Similarly, Nadia Abu-Zahra makes a powerful case for the religious knowledge of ordinary Egyptian women who visit the shrine of al-Sayyida Zaynab in Cairo.[59] Religiousness as an analytical notion may very well be mostly in the eye of the beholder.

Several anthropologists have combined advanced knowledge of the research language and textual tradition with extended fieldwork to produce sophisticated ethnographic studies of Islamic praxis. John Bowen's *Muslims through Discourse* examines stories and exorcisms recorded during fieldwork among the Gayo of highland Sumatra. The local texts come alive through explication of a local context observed and discussed at length with the Gayo themselves. The "point of departure" for Bowen's analysis is "how written texts and oral traditions are produced, read, and reread."[60] As Michael Lambek, who studied an African Islamic discourse on sorcery and spirit possession, phrases it:"The meaning of the texts that concerns us lies not in what was written into them but in what the villagers of Mayotte read out of them and what they do with them."[61] Or, consider Fadwa El Guindi's observation:"Islamic text, far from remaining frozen in Islamic scholars' specialized teaching and writings, spreads to ordinary folk through forums of collective worship and public media, and is transmitted through socialization and by oral tradition."[62] Text can be successfully wedded to context in anthropological discourse.

Perhaps nowhere has the combination of ethnography and textual analysis been more successful than among those who have done ethnographic research in Yemen. Brinkley Messick's *The Calligraphic State* affords an eloquent counter to the anthropological essentialisms to be found in the texts of Geertz, Gellner, Mernissi, and Ahmed. The introduction of this award-winning ethnography provides an enviable model for an anthropologically informed analysis of an islam defining itself as Islam:

> This book examines the changing relation between writing and author-ity in a Muslim society. Its backdrop is the end of an era of reed pens and personal seals, of handwritten books and professional copyists, of lesson circles in mosques and knowledge recited from memory, or court judg-ments on lengthy scrolls and scribes toiling behind slant-topped desks. As understood here, the calligraphic state was both a political entity and a discursive condition. My aims are to reconstruct one such textual polity and detail its gradual transformation in recent times.[63]

Informed rather than driven by theoretical currents within and without the discipline, Messick builds a thick description from the ground up. The result, consciously a construct rather than a supposed mirror image, addresses the concern of Talal Asad that observable Islamic praxis be linked both to a discursive past and an imaginable future.

A second kind of anthropological assessment from Yemen is the joint venture of ethnographer Paul Dresch and historian Bernard Haykel on the relation between Islamists and tribesfolk during the 1994 civil strife.[64] Probing the context of political slogans and participating in conversations with local activists, the authors combine their respective skills to examine how a major religious party represents itself and is stereotyped by others. The political rhetoric emanating from all sides appealed through both Quranic references and the respected medium of popular tribal poetry. Combing for clues rather than pigeon-holing into types, Dresch and Haykel turn a discrete set of events into a lesson on how Yemenis redefine their Islamic identity and ethnic loyalties at one and the same time. Where some observers might see another "fundamentalist" group, the on-the-ground negotiation of meaning defies the reductive pairing of secularism versus Islamism.[65]

For a third example of how ethnographically derived knowledge can be integrated with the historical analysis of Islamic texts, I turn to my own study of the formal genealogical model of the prophet Muhammad.[66] There is no dearth of textual and recorded oral data on Arab genealogy, because this has been such a practical passion in Arab societies and past anthropological scholarship. Muslim scholars attempting to categorize the segmentary nature of tribes often referred to the specific genealogy of Muhammad, more or less agreed upon across sectarian lines, as a paradigm for how tribal segments nested. Thus, the eleventh-century Ibn 'Abd al-Barr categorized Arab tribal structure into seven divisions, each linked in genealogical depth with a reputed ancestor of Muhammad (table E.1).

Table E.1 The tribal paradigm related to the genealogy of Muhammad according to Ibn 'Abd al-Barr

Division	Body part	Ancestor
sha'b	Skull suture	Mudar
qabila	Skull bones	Kinana
'imara	Breast	Quraysh
batn	Belly	'Abd Manaf
fakhidh	Thigh	Hashim
fasila	Lower leg	'Abd al-Muttalib

From the broadest sense of a people, a possible rendering of the Arabic term *sha'b*, to the fundamental household unit of an extended family (*fasila*), the key defining factor is genealogical depth. In contemporary ethnographic contexts, where this particular linguistic scheme is not replicated but only approximated, the inevitably political processes of fission and fusion result in a genealogy of dubious historical accuracy, but nonetheless of important influence on group association.

The formal patrilineal descent line of Muhammad back to the legendary 'Adnan, the ancestral founder of the so-called Northern Arabs, is generally accepted as consisting of twenty-two generations. Were this a complete historical record, the real 'Adnan would have lived just before the time of Christ, a timing clearly at odds with the mythological underpinnings shared by Islamic sources with Christian and Jewish texts.[67] As an anthropologist, my interest is not in the historicity of the genealogy but rather its credibility as a charter for social action in an Arab tribe. Were the Muslim scholars attempting to explain tribal structure reflecting actual practice at the time or filling in names for some other purpose? In the model of 'Abd al-Barr the sequence begins with Muhammad's grandfather as the head of the family unit; this is certainly a common structural placement in the ethnographic documentation. But the next three segment levels up are in fact linked with the next three ancestors back from Muhammad's grandfather. As a model for tribal interaction this is not at all viable.[68] Tribal sections are not created with every descending generation, as the Arab scholars surely knew. Were an ethnographer to encounter such a model in the head of an informant it would be deemed impractical, unsupportable by the evidence on the ground or in historical documentation. Such analysis does not explain what purpose the genealogy served, but it does suggest that it must be read as more than a presumed charter for tribal structure.

How Should Anthropologists Understand Islam in Light of the Postmodern and Post-Colonial Critique of the "Orientalist" Framework Embedded in Western Societies?

Thus, the traffic in symbols must be viewed not only as an Islamic phenomenon but as the product and condition of much broader and more complex historical processes, including, of course, colonialism.[69]

Michael Lambek

For the Muslims I know, and in much Islamic writing, Islam is One and to suggest otherwise is blasphemy. Does the variety of practices that the

anthropologist comes in contact with justify the omniscient observer's
perspective, or is that another form of cultural imperialism?[70]

Carol Delaney

The fact that two of the anthropological texts chosen for this study
are written by Muslims is a poignant reminder that the anthropolo-
gizing of Islam is not simply a process of othering. Western analysis
of Islam has been plagued from the start by a competitive religious
ideology, political and economic rivalry and the fact that by the nineteenth century European imperialism had come to dominate much of
the Islamic world, especially the Middle East. In his influential
Orientalism and later in *Covering Islam*, Edward Said argued that a
hegemonic discourse of Orientalism pervaded academic and popular
discourse to such an extent that Muslims and other Orientals were
not allowed permission to narrate their own stories or determine
their own destinies. In the following decades the loosely defined fields
of postcolonial and subaltern studies elaborated on the implications
of European colonial power and its neocolonial rebirth. Whether or
not it was discourse, economic power, or political ideology that
should be blamed for victimizing Muslim societies, it is seldom
doubted today that the study of Islam still suffers from ethnocentric
and racial stereotypes that infiltrate even the most avowedly objective
of published studies.

For some cultural critics, there is a lingering postmodern suspicion
that scholars cannot overcome the embedded ethnocentrism and
racism brought to Islam as outsiders.[71] If objectivity is to be defined
only as virginity, then the possibility of a neutral, nonprejudiced interpretation of Islam from the outside is rightfully suspect from the start.
None of us is without sin, which is why casting aspersion stones
willy nilly is such folly. Given the continuing tensions between various
Muslim groups and Western political and cultural intrusion, the impact
of Islam as played out in the current islams cannot be ignored. But the
abstract "can we ever be objective" malaise is not conducive to those of
us who think it worthwhile to attempt reducing tension and promoting tolerance of diverse worldviews. Advocates of the "clash of civilizations" thesis are equally unproductive, as long as the artificial division
of the world into Cains and Abels satisfies the baser instincts of polemicists hellbent on clashing. In the final analysis, there is the less politically correct but more pragmatic matter of whether we really want to
be objective in the tried and tired positivist sense. As long as influential
people justify violence by Islam, it is worth studying how and why they
do this. Caring about how religion seemingly informs destructive
behavior is a moral imperative worth keeping.

As anthropologists who conduct research in parts of the world where Western colonies have been established and unwanted cultural influence is an ongoing factor, it will not do to stop with the ethnographic present. Islam, as isolated into a conceptual black hole by Geertz and Gellner, is a discursive tradition shaped by the political events of the past several centuries. Muslims today act as they do, not simply because they follow a religion called Islam, but because their identity as Muslims necessarily responds to the social, economic, and political changes forced by Western contact. Muslims do more than keep the five pillars and read the Quran. Today they may eat at the local McDonalds in Cairo and listen to Michael Jackson's greatest hits, but they can also watch al-Jazeera cable video of Israeli tanks battlling Palestinian youth throwing stones and bombs glowing over Baghdad. "The United States may not be fully cognizant of the impact it has in provoking extremist reactions when it employs heavy-handed methods against Muslims," argues Fadwa El Guindi,"nor is there a willingness to acknowledge the corrosive effect that Israel has on the region and Islamic politics."[72] If it is true that we now live in a global village, it is closer to what one finds in Disney World than the Bethlehem visited by the wise men. Ethnography, no more than any other study in the social sciences, cannot be neutralized from the unpleasant realities of human culture at the start of the twenty-first century.

For Americans, the reality of political violence with a religious voice hit home on September 11, 2001.[73] This was not the first time acts of terrorism in the United States were justified by Muslim extremist calls for a "holy war," but the media attention and subsequent military retaliation catapulted the obscure Arabic term *al-Qaida* into instant recognition. Having conducted ethnographic research among Muslims overseas, a number of anthropologists contributed to the debate and process of understanding why suicide bombers would target civilians and icons of American superpower status. Those who had worked in Afghanistan, such as Jon Anderson, Bill Beeman, David Edwards, and Nazif Shahrani, wrote op-ed pieces and lectured in various forums.[74] Others used their knowledge gained living abroad to discuss the portrayal of Islamic radicalism in Egyptian cinema,[75] local responses to civil violence in Indonesia,[76] and comparison between the terrorism of al-Qaida and the Sicilian mafia.[77] Andrew Shryock reflected on the local impact of September 11 on the Arab Muslim community he knew in the Detroit area.[78] A small oasis of ethnography could be found among the plethora of overexposed media pundits.

Beyond ideology but not quite beyond theology, modern scholars who study Islam need to address their own institutionalized fears of the politics of religious intolerance. As John Bowen asks, is "radical political Islam," like Afrocentrism and Zionism "beyond a certain anthropological pale?"[79] The tact Bowen takes is a civil one, that "we" in our pluralistic confession of secular democracy find it hard to be sympathetic to "state-enforced religious law." One obvious reason for this is our reading of European society, where the meshing of religion and civil authority has a bloody history. Is it any wonder that academics, no less than the public at large, have problems dealing with recent "religious" conflicts, no matter what the nonreligious factors acknowledged, that involve angry Muslims in Palestine, Afghanistan, Iran, Sudan, Nigeria, India, Indonesia, the Balkans, and Chechnya? Having academically assigned political ideologies of Nazism, Fascism and Communism to history, it now seems to be the case that intolerance is relegated to feeding on religion. Bowen also notes a more nagging personal dilemma in which we are less willing to accommodate "religion" to the extent it has more of an impact in our own lives. It is a lot easier to write about saints and sufis than to come to terms with Hamas suicide bombers.

As I write this epilogue, and no doubt for the foreseeable future, covering Islam in American society is fixated by clash talking.[80] Less than a decade before the attack on the twin towers, historian Samuel Huntington provided the rubric for the current representation of Islam as prone to ideological terrorism. Speculating on the state of the world after the meltdown of the post–World War II Cold War, Huntington spoke of a "clash of civilizations," in which the West had entered a high-stakes military standoff against the rest.[81] The primary rivals of the West were redefined not as nation-states but religiously dominated civilizations, the oldest enmity being with Islam. This was hardly a new thesis, but anthropologists have been tearing away at such ethnocentric hubris for decades. Those who look at world events through this narrow lens of us versus them need an essentialized Islam to hook onto terrorism. Here is where ethnography can help mitigate our collective fears, as Robert Hefner shows in his perceptive analysis of recent Indonesian politics and Jonah Blank documents for the digital-generation Daudi Bohra in *Mullahs on the Mainframe*.[82] Muslim society can be imagined as being "civil" and capable of achieving balance on its own terms.

Should there be an Explicitly Anthropological Definition of "Islam"?

The anthropological study of Islam is one that has been plagued by problems of definition. What exactly are we studying? Local practices, universal texts and standards of practice, or something else entirely?[83]

Ronald A. Lukens-Bull

At the beginning of this book I re-envisioned the poignant parable of the blind Hindu and the elephant in order to suggest that the idea of an elephant depends very much on the bodily parts a mere human is able to reach. Elephants, for the pragmatic realism most of us follow by default, obviously exist outside the mental constructs of Ivory Towering hermeneuticists. The question remains as to whether or not we as scholars are inevitably blind to the whole, destined to the fatalistic flaw of rendering what we assume must exist by what we are able to grasp in some kind of disciplined contact with the beast. If all meta-narratives are false, as the seemingly logical conclusion of postmodern meta-theory suggests, then we are forced to admit our blindness to meaningful wholes, whether we choose to swallow Derrida whole or dance warily with Lyotard. Thinking ourselves into a corner with that elephant, would there then be an option to simply running our hands—God help the poor soul who brings up the rear—over the surface near at hand. Islam is not an elephant and scholars are hopefully not content to be blind keepers. But the paradox survives the absurdity of the parable. No matter how anyone tries to define Islam, it will always be one among many representable options.

As an anthropologist primarily concerned with how Muslims act and view their world, I agree with Abdul Hamid el-Zein that a pre-given, ideal-typed and essentialized idea of Islam has little heuristic value as an anthropological concept. What does make analytical sense is considering how such notions and definitions influence behavior and flavor the full gamut of social interaction, institutions, and socially relevant discourses. I am perfectly satisfied to work from the indigenous notions of Muslims about the meaning of their own faith and to learn from alternative models proposed by Marxists, feminists, pragmatic political scientists, and even the occasional scholar claiming to be totally objective. Unfortunately, the definitions advocated consciously or in default by Geertz, Gellner, Mernissi, and Ahmed are not "anthropological" in the sense of applying across cultures to any "religion." They are all, as Talal Asad reminds us, influenced by "a certain narrative relation" that is ultimately either supportive or oppositional to how Muslims define themselves.[84] Searching for the idea of an anthropology of Islam, I argue, should not lead us beyond ideology and theology but rather probe these very powerful discursive traditions through thick description of ethnographic contexts. Observing Muslims in particular "islams" is one of the few things that anthropologists have been able to contribute to the broader academic interest in how Islam is continually defined and redefined and, indeed, how religion itself is conceptualized.

In critiquing some of the seminal texts running with a specifically anthropological approach to Islam, I do not support the nihilistic cop-out that Islam does not exist in a positive, effectual, and historically documentable way. Blind men can hear an elephant stampede and get properly mauled for their philosophical skepticism. Muslims will continue to be Muslims and many new converts will be found despite any attempt by academic scholars to explain the process away. Many scholars have a tendency to reduce what Muslims see as a vibrant and revelatory faith to something not essentially religious at all. For example, Pierre Bourdieu in his influential studies on North African society writes as if Islam were primarily a mislabeling of the local *habitas*. "Islam" is absent from his *The Logic of Practice*, except for its ultimate appearance under the index entry for magic.[85] In pointing out this kind of omission, Carol Delaney raises an important issue: "We must ask whether an approach to the study of human culture that is grounded on universalistic premises about work, the division of labor, and the transformation of society is appropriate for all cultures, including Muslim cultures."[86] Indeed we must, although it is hard to imagine an anthropological response that would not ultimately insist that what is true about Muslims must be potentially true in any other human context. Individual cultures must be appreciated for their individuality and not unduly typed into artificial and unproductive categories, but if anthropology is not about a pan-human, and at times pan-primate, sharing of something called "culture" as such, what do we as observers of human diversity have to contribute beyond encyclopedic data banking for the polemicists?

Defining Islam will not explain what Muslims do and why they do things differently over time and space. The same should be said for Christianity, Judaism, Hinduism, Buddhism, or any of the religions readily defined in introductory texts and in popular culture. The real issue is how Islam, however defined, is represented in native or indigenous views and by outsiders. The anthropologist has the opportunity to be ethnographically present to observe what Muslims do and say. "What is distinctive about modern anthropology," comments Talal Asad, "is the comparison of embedded concepts (representations) between societies differently located in time or space."[87] Anyone can compare concepts; anthropologists try to find out what people in various social contexts think and do with those concepts. Our best comparison is applied ethnography, which can be an important corrective to armchair philosophy.

I do not need or even desire an anthropological definition of Islam, especially an essentialized model that inevitably fudges the observable variations in Muslim behavior and thinking. It is enough to start from the definitions that are useful to understand human behavior as such. This is hardly a new idea. In a significant but rarely consulted essay about anthropological perspectives on Islam, Jean-Pierre Digard argued that instead of an anthropology of religion, including the idea of an anthropology of Islam, anthropologists should place religion in its economic, political, and social contexts. "Je suis convaincu," argues Digard, "qu'il avait raison et que, si l'anthropologie peut apporter une contribution spécifique à la connaissance des religions (et donc de l'islam), c'est précisément en les replaçant dans leur contexte obligé que constituent l'économie, l'organisation sociale, etc. des peuples qui les pratiquent."[88] The key here, reflecting my bias as someone trained profitably in American anthropology, is the utility of a concept of culture. Definitions of culture are notoriously more contested than those of a specific religious tradition like Islam. British anthropologists like Gellner have a skeptical view of the American interest in culture, preferring to stick with the more material economic and political aspects of society as an institutional base; most do not wax and wane as philosophically as Gellner. But just as theologians have to deal with the idea of "God," whether they believe in it or not, anthropologists inevitably must go beyond the ethnographic context observed to a broader comparative understanding of how every given human act relates to the potential for specifically human interaction.

So much has been written about "culture," capitalized or not, that the very idea of bringing it up in the closing paragraph of a book about the religion of Islam would seem to border on the absurd. Anthropology, as I understand and practice it, is about culture or it is nothing.[89] Call it what you will, redefine it as you should, when I study Muslim behavior I see not only individual others negotiating their own identities in an ever-changing world of options; there but for the grace of genealogy go I. Or, there but for the uncontrollable fact of historical choice, goes anyone whose ancestors have been evolving over several million years into a species we arbitrarily but sympathetically call human. Anthropology can only explore what it means to be Muslim against a shared humanity revealed by the always tentative, but not easily ignored, findings of modern science and challenging reflections of critical philosophy. Beyond that, in the realm where ideology and theology reign supreme, anthropology has little to contribute. Studying what Muslims believe or fail to believe may say something about human nature, but it offers no window into the truth of revelation. The anthropologist observes Muslims in order to represent their representations; only Muslims can observe Islam.

Notes

Introduction Anthropology and Islam

1. Asad (1986b:1).
2. Rahman (1966:xx). A similar sentiment can be found in Omid Safi's (2003:8) description of progressive Islam: "It cannot survive as a graft of Secular Humanism onto the tree of Islam, but must emerge from within that very entity."
3. el-Zein (1977:242). As Mark Woodward (1996:6) advises, non-Muslim scholars "can describe, but not define, Islam." If historical analysis is confused with the quest for religious truth, adds Woodward, "both endeavors are in peril."
4. It is relevant to note that Ahmed (1986:187–188) refers to Geertz and Gellner as "two of the most prominent Western anthropologists and both leading their distinct schools of anthropology on either side of the Atlantic . . ." Raymond Firth (1981:597) adds: "Two thoughtful general studies of Islam, fairly sharply contrasted, are by Clifford Geertz (1968) and Gellner (1981)." Graham (1993:504, note 14) also quotes Geertz and Gellner in the same breath.
5. Redfield (1960:40). In retrospect, it is highly suspect that any so-called isolated culture was ever really isolated, except in the short term. Redfield's point is that in some contexts it is impossible to understand village dynamics without direct reference to the wider world in which it is contextualized.
6. Redfield (1960:40).
7. von Grunebaum (1955:28). I say this not to belittle the historian, but to qualify the prevailing sentiments at the time. As Launay (1992:4) states, "It was the task of the 'Islamicist' to describe Islam, and the task of the anthropologist to describe its 'influence' at the 'periphery' of the Muslim world, if not its local peculiarities within the Muslim 'core.'"
8. Redfield (1960:49). By "Islamist" Redfield means a scholar of Islam, not a "fundamentalist."
9. Redfield (1960:57).
10. See Digard (1978:501).
11. I view the limitations in Redfield's argument more as a general paradigmatic blindness across disciplines rather than any personal fault. It

would be disingenuous to assert that Redfield necessarily saw the "great" as being essentially "better" or that he was unable to appreciate that small was indeed beautiful.

12. Laroui (1976:44–80).
13. See especially Asad (1986b:6); Bowen (1993a:185–186); Eickelman and Anderson (1999:12); Eickelman (2002:243–244); el-Aswad (2002:5,17–18; 2000); El Guindi (1999:xiv); Lukens-Bull (1999:4); and Tapper and Tapper (1987:70). One anthropologist not quite ready to abandon the dichotomy is Ernest Gellner (1981:4–5), who asserts that only the "Great Tradition [capitals in the original] is modernisable." To make things interesting, Gellner proceeds to argue that the "old Great Tradition became the folk version under modern conditions, which also made that folk far more numerous and far more weighty in the state." So the Great may not be so great and the Little not so folksy.
14. el-Zein (1977:246).
15. Richard Martin (1996:206): "The great tradition is reflected in the more readily accessible literary products of theologians, judges, and lawyers of the Islamic academics of learning. The little tradition exists in the tribes, villages, and groups within cities where, in the process of social organization people must attempt to apply the Quran and the Sunna to everyday life, in varying contexts and under diverse circumstances." See also Graham (1993:498, note 7). This distinction obviously works, but it no longer reflects the changing perspective among most anthropologists to treat culture as a dynamic interplay not easily reduced to highs and lows.
16. This includes widely distributed public statements by Jerry Falwell and Franklin Graham. Symptomatic of the politicization of the debate is the attempt to stop the University of North Carolina in 2002 from using Michael Sells' acclaimed *Approaching the Qur'an* (1999) as summer reading for incoming freshman. As Sells (2002) noted with irony, "In effect the plaintiffs are suing the Koran on behalf of the Bible." Unfortunately there appears to be no statute of limitations on religious intolerance.
17. Lewis (2002:159).
18. Kramer (2001:47).
19. From the translation by Dawud (1968).
20. Quoted in Southern (1962:31).
21. Makdisi (1989:175). The Arabic is *ijazat al-tadris* and it was used primarily to ensure proper credentials for teaching about the faith.
22. Smith (1957:28). Among anthropologists who criticize this approach are Talal Asad (1986b:15) and Nadia Abu-Zahra (1997:37). As Gregory Starrett (1995:964) explains, "The persistent claim that Islam is a religion of 'orthopraxy', concerned with correct performance of ritual, rather than of orthodoxy, concern for correct belief, has been used in part by Western scholars to distance outer-directed Islam from inner-directed Christianity, the implication being that 'their' religion is empty while 'ours' has intellectual content."

23. For a recent example, see Esack (1999) or the older classic by Fazlur Rahman (1968).

24. This was the mid-nineteenth-century view of Ernest Renan, elaborated more recently by Chelhod (1958). Most scholars now see beyond the desert vs. sown reductionism inherent in this approach; see Digard (1978:506–507) and Donner (1981).

25. Norton (1924:389).

26. The term ethnology has not survived in American usage, where it came to be associated with armchair analysis not unlike the folkloric compilation of Sir James Frazer or the diffusionist Kulturkreise theory of early twentieth-century German anthropologists like Vater Schmidt. Schmidt, a devout Catholic, wrote in his *Der Ursprung der Gottesidee* (1926ff) that humans in their most primitive state were closest to the supreme being. Citing this, a Muslim writer in 1940 could claim that the results of modern ethnological investigation proved the truth of Islam because it had determined the first religion was monotheism (Ehrenfels 1940:446).

27. It is interesting to note that Peters (1966), who had already conducted fieldwork among Cyrenaican Bedouin, viewed the nineteenth-century Semiticist and theologian, William Robertson Smith, as a kind of patron saint for the social anthropology of Middle East kinship.

28. Zammito (2002).

29. Brantlinger (1990:159).

30. Said (1989:225). I respond to this charge in Varisco (2004).

31. The fact that I assume there is no need to cite the first names of these iconic theorists should be taken as suggestive in and of itself.

32. Handleman (1995:343).

33. See Moseley (1983:104–110).

34. Van-Lennep (1875:5).

35. Sweet (1970).

36. Roy Ellen (1983:54) regards Hurgronje as "an ethnographer in the grand manner; insightful and analytic as well as systematic and comprehensive." There is no denying the breadth of Hurgronje's research, but it was still Orientalist or armchair in approach rather than based on participant observation or the critical use of social science methodology. Edward Westermarck (1862–1939) has been cited as "a pioneer of local anthropological fieldwork," by Timothy Stroup (cited in Eickelman 2002:43), even though he only visited Morocco intermittently over a number of years.

37. Evans-Pritchard (1949). As a military officer during the war, the context is somewhat unusual. The focus of this text is on the social structure of the Bedouin tribes and the political role of the Sanusi religious order. Unfortunately, there is very little information about how Islam is practiced. Similarly, Emers Peters studied the Cyrenaican Bedouin after World War II, but ignores the role of religion as such. In a reflective article Peters (1984) later remarks on the failure of ethnographers to discuss ritual among Muslim pastoralists.

38. Wolf (1951). Anthropological critiques of Wolf have been given by Lagrace (1957) and Eickelman (1967); see also Asad (1980).
39. Aswad (1970). It is telling that Aswad's article appears in a volume of readings edited by an anthropologist in which the two major selections on Islam are by historians Anne Lambton and Bernard Lewis.
40. For his approach to Islamologie, see Chelhod (1969).
41. Gilsenen (1990:225), whose *Recognizing Islam*, first published in 1982, is one of the best anthropological studies on the broad theme of Islam in practice.
42. The third edition of Coon's popular text was published in 1962.
43. Joseph Chelhod wrote on Arab tribal structure, primarily based however, on Arabic sources and the travel literature.
44. Antoun (1976:159–166). In this regard it is useful to observe that articles devoted to Islamic practice in ethnographic contexts were often published in non-anthropological journals before the mid-70s (e.g., Barclay 1963; Salzman 1975).
45. el-Zein (1977).
46. In the following discussion I do not attempt a comprehensive survey of anthropological studies relevant to Islam. For bibliographical help, see Eickelman (2002:306–311); and Strijp (1992).
47. For Iran, see (Barth 1961); for Pakistan, Ahmed (1976); for Turkey, Bates (1973); for Saudi Arabia, Cole (1975); for Israel, Marx (1967).
48. See Ammar (1954) and Antoun (1972). Examples of village ethnographies by nonnative speakers include Barclay (1964); Duvignaud (1970); Magnarella (1974); Stirling (1965); Sweet (1974). My point is not that these anthropologists were unable to communicate in the field, but that no attention was paid to literary sources, including the Quran. It is not surprising that "Islam" is subsumed under "The Symbolic Order" in the early review of Middle East ethnography by Fernea and Malarkey (1975:193).
49. Sweet (1974:220–224).
50. Stirling (1965:274–275).
51. For Algeria, see Bourdieu (1960); for Afghanistan, see Dupree (1973); for the Hindu Kush, see Canfield (1973).
52. Gellner (1970).
53. Crapanzano (1973).
54. Eickelman (1976).
55. Lewis (1984) and Gilsenen (1973).
56. el-Zein (1974).
57. el-Zein (1974:221–280).
58. For pilgrimage, see Delaney (1990); Eickelman and Piscatori (1990); Fischer and Abedi (1990:150–221); and Young (1993). For prayer, see Bowen (1989); and Mahmoud (2001b). For Egyptian *mawlid*, see Abu-Zahra (1997:205–230) and for Turkish *mevlûd*, see Tapper and Tapper (1987); and the critique by Abu-Zahra (1997:41–49). Mortuary rituals in Egypt and Tunisia are discussed by Abu-Zahra (1997:49071). For sacrifice, see Bowen (1992); and Hammoudi

(1993). Antoun (1968) discusses traditional Ramadan fasting, while Armbrust (2000) contextualizes a popular Ramadan television show in Egypt. Hefner (1985:104–125) examines regional variations in the Javanese *slametan* ritual.

59. el-Aswad (2002).
60. Messick (1993) and Varisco (1989, 1994).
61. See especially Banks (1990); Edwards (1996, 2002); El Guindi (1981); Gladney (1999); Hefner (2000); Hirschkind (2001); Houston (2001); Munson (1987); Nagata (1982); Shahrani and Canfield (1984); Shamsul (1997); Tapper and Tapper (1987); and Toth (2003). For Iran, see Fischer (1980) and Swenson (1985). For Turkey, see White (2002). The *American Anthropologist* included a special section called "In Focus: September 11, 2001" with analysis of the "Islamic" dimensions of the terrorist issue by Andriolo (2002); Hefner (2002); Mamdani (2002) and Varisco (2002b) among others.
62. See Ahmed (1992:169–177); Rosen (2002:158–173); and Werbner (1996).
63. Eickelman and Anderson (1998:16).
64. Bowen (1997).
65. Edwards (1995); Starrett (1995, 1998); see also Eickelman and Anderson (1997).
66. See Abu-Lughod (1993); Adra (1997); Armbrust (1996, 2002); Gordon (1998); Murphy (2000); White (1999).
67. Anderson (1999); see also Varisco (2000, 2002b).
68. I have chosen to place "medieval" in parenthesis because I find it unduly ethnocentric for periodization of Islamic history.
69. Geertz (1968:vi).
70. Gellner (1973:191).
71. Gilsenen (1982:9) remarks that his "own experience of Islam began with a surprised and uncomfortable recognition that things are not what they seem." This is not unique to anthropologists, by any means.

Chapter 1 Clifford Geertz: *Islam Observed* Again

1. Moosa (2003:115).
2. Said (1979:326). Woodward (1996:29) contrasts Geertz's Weberian sociological approach with Orientalist categories, but surely Said would consider Weber's views on Islam as Orientalist. Lukens-Bull (1999:2) notes that Said overlooks Geertz's "*latent* Orientalism." After Geertz's less than enthusiastic response to *Orientalism*, Said (1982:44–46) criticizes Geertz's critique of his own work as "rather trivial," and implies that Geertz sees him as a "wog." Still later, Said (1985:5) disses Geertz for his "standard disciplinary rationalisations and self-congratulatory clichés about hermeneutic circles" as a foil for Johannes Fabian's "unique and important" *Time and the Other*.

3. el-Zein (1977:227).
4. Launay (1992:1).
5. Munson (1993:182).
6. As noted by Sewell (1997:35), among others. The literature about Geertz is extensive. No post–Marvin-Harrris history of anthropology has failed to account for his influence. Rare is the anthropologist who has not commented in print on Geertz. Barrett (1996:158), for example, comments: "In recent years he has nudged Lévi-Strauss aside as the discipline's reigning genius." James Boon (1982:140) calls the Geertzian approach "pragmatism, to have a tail to wag it with." For the influence of Geertz on social historians, see Walters (1980) and Spiegel (1997:11–13). Mine is not the initial replay on Geertz's own voluminous wordplay; see Handleman (1995:343) for a poetic view of how Geertz represents himself on a book cover.
7. Geertz (1968:viii); see Geertz (1966) for his own genealogical charting. The theoretical roots of Geertzian anthropology have been discussed widely, e.g. Martin (1984) and Trencher (2000:33–46). The Weberian strain in Geertz is usually taken for granted; Firth (1969:909) goes so far as to assert *Islam Observed* "is the kind of book Max Weber might have written had he studied oriental societies at first hand instead of from the literature." Of course, then he would not have been Max Weber. On the other hand, Laitin (1978:589) accuses Geertz of not being "fully in the spirit of the Weberian tradition" and Siegel (1999:62) contends that Geertz does not share Weber's sense of meaning. It is thus not always the case that one man's Weber is another's Geertz.
8. Martin (1984:19).
9. Geertz (1968:v).
10. Geertz (1968:v).
11. Geertz (1968:62).
12. Geertz (1968:116).
13. Geertz (1968:vii).
14. Psychologically, failing to provide specific bibliographic references in the narrative for points derived from books and articles may make the published essays easier to read, but this also dulls the reader to the amount of material actually "borrowed," to use Geertz's (1968:v) own spin.
15. In his ethnographic monograph on Java, Geertz (1960:ix) begins by noting that "the research project upon which this work is based has extended over a six-year period" but later in an appendix (pp. 383–386) we learn that about 18 months were actually spent in fieldwork.
16. Woodward (1996:31). Geertz does not get around to discussing Islam as such in Indonesia until chapter 10 (of some 22) in the *Religion of Java*; his generic description is derived primarily from Gibb and Snouck Hurgronje.

17. His main effort was directed at economic activity in a Moroccan bazaar (Geertz 1979).
18. Geertz (1968:vi).
19. So much has written about this critique, that it seems redundant to cite references. Trencher (2000) provides a survey that focuses on ethnographic texts written about Morocco. The neo-canonical articles in Clifford and Marcus (1986) are a good a starting point. For a particularly eloquent summary of the problem, consider Lambek (1993:27): "Fieldwork is an encounter, a hesitant grappling of epistemological horns. If I were much the stronger, the Other would hold no interest for me; if the Other were much the stronger, I would have no independent perspective from which to report. What we create if we are both successful and interested is a mutually comprehensible dialogue, a fusion of horizons, the ground for further conversation, not a unified theory. The potential for self-deception is of course very high."
20. Rabinow (1977:3).
21. Gupta and Ferguson (1997:1) express this well: "Indeed, we would suggest that the single most significant factor determining whether a piece of research will be accepted as (that magical word) 'anthropological' is the extent to which it depends on experience 'in the field.' "
22. Geertz (1973:412–453). This book was reprinted, but not revised, in 2000.
23. Crapanzano (1986:74).
24. Clifford (1988:41).
25. Geertz (1968:vii).
26. Geertz (1968:vii).
27. Geertz (1968:viii).
28. "In cultural studies," suggests Adam Kuper (1999:119), "he has become a guru for the *les marxisants* practitioners." Geertz is aligned with postmodernism, especially for his influence on the new historicism, by Stuart Sim (2001:254–255) in *The Routledge Companion to Postmodernism*.
29. Benda (1962:406). As a graduate student, my initial attraction to Geertz was quite honestly the brilliance of his prose and the verve of his wit. I still maintain that "Deep Play," the cockfight essay dismissed by Crapanzano, is so full of puns, erotic as well as exotic, that it should be required reading of any scholar with an intact, preferably untactful, sense of humor.
30. Greenblatt (1994:97). Bauerlein (1997:30) suggests that "Cultural poetics takes from Geertz a set of motives and premises, not arguments and demonstrations."
31. The phrase is from Asad (1986b:8).
32. Geertz (1968:1). Obviously Geertz is not referring to the problem addressed in his influential essay on religion as a cultural system.
33. Handleman (1994:345–346) refers to Geertz's style in essays as "unsettling, subversive," adding: "While floating along the stream of

words—taking that kind of trip—one is shockingly caught in a vortex, the stream flowing into a whirlpool that sucks the reader in, down, spitting him out elsewhere (and elsewhen)—it's become that kind of trip: the kind that reacts recursively against the reader's attempts to fix and stabilize the text." This is a powerful metaphor, but I think it applies equally well to the intellectual laziness of readers who are swept off their feet as it does to the author. If more anthropologist authors had such dazzling rhetorical power, perhaps the discipline would not be as marginalized.

34. Geertz (1968:2).

35. Geertz (1968:v).

36. Geertz (1968:vi). Abaza and Stauth (1988:348) take Geertz to task for sketching a comparison of Morocco and Indonesia "without giving close reference to the interplay between class and religious style of life practice." In fairness, *Islam Observed* was published as a series of lectures rather than an ethnographic text. The problem is that many scholars read this sketch but do not take time to examine the oils and installations in the gallery of modern ethnography.

37. Geertz (1968:104) does not appear to be entirely in sympathy with the primary goal of the lecture series to pacify the ongoing warfare between science and religion. Such warfare, which he likens to "a succession of random skirmishes, brief, confused, and indecisive" is "not only not over; it is quite likely never going to end."

38. The extent to which Geertz does "science" is an issue of considerable debate by his critics. Perhaps, it is best to agree with the modest statement by Shankman (1984:263) that this is "a science with a difference."

39. Benda (1962:405) makes a similar criticism of *The Religion of Java*, in which a vast number of original texts and derivative studies have been ignored; the question remains: "Why approach his subject as if it were a *tabula rasa*?" As Nissim-Sabat (1987:937) observes, in *Islam Observed* Geertz "tells us nothing about what Muslims preach, believe, or practice. What is discussed is their style, and who's to judge?" Hefner (1997:14) notes that by ignoring the available scholarship on Islamic texts, Geertz follows a "narrow standard for distinguishing what is and what is not Islam." Specific flaws in Geertz's failure to consult Javanese primary texts are enumerated by Woodward (1989:245–247).

40. Geertz (1968:19). Munson (1993:x), in his critique of Geertz, provides "a less ethereal version of Geertz's conception of the 'social history of the imagination' (1968:19)."

41. Benda (1962:405), in an ultimately favorable review, claims that Geertz "underestimated, and indeed almost ignored" the role of mysticism in Java. Marshall Hodgson (1974(2):551, note 2) calls Geertz's tendency to label suspect Javanese Muslim customs as Hindu-Buddhist a "major systematic error" of *The Religion of Java*. For an earlier critique, see Woodward (1989). Suggesting that Geertz's views

on the syncretistic dimension of Islam in Java have been distorted, Hefner (2000:xix, 28, 232, note 3) comments that *The Religion of Java* "shows a considerable understanding of Javanese Islam"; although he takes issue with some of the points raised in *Islam Observed* and *Peddlers and Princes*. Lukens-Bull (1999:8) takes exception to Geertz's distinction between mystical and normative Islam.

42. As Boon (1982:108) observes, in Geertz's interpretation of Islam "ethnographic particularities themselves emerge as subtle, operational ideal-types." Firth (1969:910) is more blunt in complaining that Geertz's ideal types are "a bit too like selected bundles of qualities."

43. The terms and phrases here are those of Geertz (1968).

44. Geertz (1968:119).

45. Geertz (1968:74).

46. Geertz (1968:35). For a critical assessment of Geertz's treatment of the Moroccan, see Munson (1993:1–34); for a critique of Geertz's discussion of Sunan, see Roff (1985:24) and Woodward (1989:97). It is worth remembering that Sunan cannot be exclusively coded as a "religious" character, since he is also assumed to be the man who introduced the central aesthetic forms of shadow theater, gamelan music, and the slametan ritual.

47. I am not claiming that Geertz preferred his experience in Indonesia to that in Morocco, as in the extreme case of Colin Turnbull's change of heart between the Mbuti and the Ik, but Geertz's rhetoric suggests that the Indonesia type is a more tolerant and rational approach.

48. Geertz (1968:106).

49. Ellen (1983:63). Ellen (1983:59) further comments: "Such broad terms have their dangers, and it is rare for the characteristics of any one to account for the ethnographic particulars of actual practice."

50. Geertz (1968:106).

51. Munson (1993:33). Binder (1988:99) adds that Geertz's use of these two legendary personalities as metaphors "does not succeed in liberating the text so much as, initially, masking its authoritative judgments."

52. As Benda (1962:403) complained about *The Religion of Java*, Geertz falls into the interpretive trap of *pars pro toto* in which a theoretical village eyeview is overgeneralized. Benda was concerned that Modjokuto, the main fieldsite in Java, was not typical of Javanese Muslim communities.

53. Kuper (1999:100).

54. Geertz (1973:90). Martin (1984:20) comments that Geertz's definition "has enjoyed enormous popularity among historians of religion." Tibi (2001:16, 20–21, 33–37, 109) is among those who adopt Geertz's approach, even if not becoming a Geertzian himself and admittedly siding with Gellner against Geertz on the universality of reason. Asad (1993:54) recommends "unpacking the comprehensive concept which he or she translates as 'religion' into heterogeneous elements according to its historical character."

55. Among the many reprints of the religion article, it is included in the recent reader by David Hicks (1999). In quoting the Geertzian definition, Klass (1995:22) calls it "what a religion is" rather than a rubric for defining religion as such. This does not stop the stream of praise that goes so far as to basically credit Geertz with helping "set the stage for the anthropology of religion" (Scupin 2000:12) as though there was no anthropology of religion before; later in the same volume Scupin lauds Geertz's argument in *Islam Observed* without detailing the extensive critique of this text.

56. Geertz (1973:89).

57. Geertz (1973:90).

58. It is not my intention to critique Geertz' interpretive approach as such, nor to argue that his definitions of culture and religion are not useful. It is certainly the case that Geertz's approach to religion has been influential for many anthropologists who conduct ethnography among Muslims, e.g., Antoun (1989:233); Delaney (1991:7); Lambek (1993:xiii); Lukens-Bull (2001); Woodward (1989:viii). Specific critiques of Geertz's approach to religion are provided by Asad (1993:27–54), originally published in 1983; Binder (1988:97–103); Munson (1986, 1993); Pals (1996:259–263); and Shankman (1984). As Jackson (1989:177) avers, "Replacing 'reason' with the notion of 'meaning,' anthropologists such as Geertz invoke hermeneutics and rhetoric to blur the distinctions between science and art, a move which, in anthropology, risks encouraging the production of bad science and bad art." Sharabi (1990:10) points out that Geertz uses "a language and style that obscure the somewhat ordinary character of his conclusions." Tibi (2001:188) argues that Geertz "regrettably overlooks the important global context pertinent to cultural analysis." For those interested in Geertz's self-defense, his more recent volumes are primary resources, including a wide-ranging interview with Richard Handler (1991).

59. It is interesting that Geertz chose not to revise his 1963 paper on religion in either its 1966 publication or later inclusion in his 1973 book. He was certainly not alone, as the article implies, in assessing religion as an anthropologist. Consider the major texts by Douglas (1966); Evans-Pritchard (1965); and Wallace (1966). The essays by Goody (1961) and Horton (1993[1960]) are cited but ignored.

60. Klass (1995:3). Dissatisfaction with the adequacy of existing approaches to religion is common, e.g., Guthrie (1980:181). Adams (1985:vii), echoing a claim he made earlier in 1967, insists that "historians of religion have failed to advance our knowledge and understanding of Islam as religion and that Islamists have failed to explain adequately Islamic religious phenomena."

61. Geertz (1973:88) quoted this phrase from an article by Janowitz (1963). In *Islam Observed*, Geertz (1968:54) asks rhetorically: "Is the comparative study of religion condemned to mindless descriptivism

and an equally mindless celebration of the unique?" It is never clear why Geertz thinks that is all there was.

62. Geertz (1973:88).

63. The phrase is quoted in Geertz (1973:89). The influence of Parsons and Shils is acknowledged in Geertz (1968:v) and analyzed by Hefner and Hoben (1991). It seems odd that Geertz felt it necessary to quote sociologists for this emphasis on culture, when "culture" was still a dominant conceptual scheme in the discipline in which he was trained.

64. Although much of the succeeding discussion is critical of the way Geertz packages his understanding of "culture," this should not be seen as wholesale movement to abandon the culture concept. As Clifford (1988:10) states, "Culture is a deeply compromised idea I cannot yet do without."

65. Geertz (1973:89). The inspiration for this definition appears to derive in part, as Asad has noted, from the synthesis by Kroeber and Kluckhohn (1952:181) that "Culture consists of patterns, explicit and implicit, of and for behavior acquired and transmitted by symbols." This compilation was reprinted in paperback in 1963.

66. Reiss (1967:22).

67. Central to the Benedict-ine doctrine of cultural determinism is the idea that "culture, like an individual, is a more or less consistent pattern of thought and action" (Benedict 1934:46). Like Geertz, Benedict was reaching out for non-anthropological theory of the day, borrowing heavily from Gestalt psychology and Spengler's historical configuring of Western civilization. This was, in her mind, thoroughly in harmony with modern science at the time, which in a now-dated sense it was.

68. I take this from the OED.

69. Crapanzano (1986:72).

70. Talal Asad (1993:30).

71. See Klass (1995:23).

72. Among the criticisms of Geertz's approach as "scientific" is the complaint voiced by D'Andrade (1995:248) that "there was no method of validation; since the meanings were not in anyone's mind, even an unconscious one, no method of verification was possible."

73. Geertz (1973:92).

74. Geertz (1973:92).

75. Geertz (1973:412–417).

76. For the former, see Geertz (1973:114–118)); for the latter, see Geertz (2000:180–184). Nissim-Sabat (1987:937) complains that the only women in Islam Observed are those in lists.

77. Biddick (1994:21). Another way of pointing out Geertz's lack of engagement with the political aspects of his theoretical spinning is Keesing's (1987:166) poignant witticism: "Where feminists and Marxists find oppression, symbolists find meaning."

78. Tylor (1871:8). Elsewhere in his discussion, Geertz (1973:100) asks not to be dismissed for holding a "Tyloreanism" in the sense of

reducing the exotic to the gentlemanly common-sense logic of the interpreter. As David Hicks (1999:11) ruefully observes: "Geertz's definition lacks any mention of spirits or the supernatural, however, and a skeptical reader might wonder what there is that is distinctively 'religious' about it."

79. Asad (1993:43).
80. Waghorne (1984:32).
81. Geertz (1973:91).
82. Geertz (1973:91).
83. Geertz (1973:93). William Sewell (1997:47), a social historian, characterizes this model of/for as an "exceptionally fruitful observation." Dan Handleman (1994:345), less sanguinely, sees it as a scholarly jingle not helpful for theory-building.
84. Geertz (1973:118).
85. Michael Lambek (1993:xiii) remarks: "In his ovular essay on religion Geertz presents an ideal model of—or perhaps for—religion; this is religion in the abstract." In a concrete sense, a cult derived from this Geertzian definition would make it a model "for."
86. It is curious that scholars continue to quote Geertz' prepositional model of models when there is an extensive literature available on what models mean. As Bloch (1953:228) cautions, in an influential reference certainly accessible to Geertz before publishing his *Interpretation of Cultures*, this use of models in an "as if" mode makes them into "heuristic fictions," where "we reap the advantages of an explanation but are exposed to the dangers of self-deception by myths."
87. Ironically, given the humanities roots of Geertz's acknowledged forebearers, critics of his interpretive mode might argue that his *Geisteswissenschaften* is heavy on the *Geist* and weak on the *Wissenschaft*. As Abaza and Stauth (1988:349) argue, Geertz's "attempts of replacing 'positivism' with hermeneutics might easily lead to the conclusion that the local truths are mere shadows of his own general vision."
88. Cornell (1998:xli) criticizes Geertz for presenting a definition which " 'poisons the well' against spiritual realities and strips religion of its claim to truth." Guthrie (1980:183) observes that Geertz "fails to distinguish religious models from other models."
89. Geertz (1973:125).
90. The various interpretations of what the Bororo meant by claiming to be parrots have been brilliantly analyzed by J. Z. Smith (1978:265–288).
91. Geertz (1973:121–122).
92. The original statement was a German traveler's rendering of what a Bororo man allegedly said in the late nineteenth century (Smith 1978:265–266). Instead of approaching the problem as a translation issue, Geertz does his best to interpret what a rational Bororo must have meant.
93. Asad (1993:45).

94. Laitin (1978:588–589), who chides Geertz for not discussing the "practical religion of Islam."
95. Geertz (1973:vi).
96. It would be better to use "Discerning Islam," as Eickelman (1985:114) devises for a section in his social biography of a Berber judge.
97. See, for example, his presentation of the discussion between Lyusi and the Sultan (Geertz 1968:33–35).
98. Geertz (1968:116–117). The other "named" individuals are historical figures or politicians, especially Sukarno and Mohammed V. Ironically, even though real Muslims do not appear in the text, Indonesian scholars use Geertz's work as proof that local Muslims have departed from the straight path of Islam (Hefner 1996:291–292; Woodward 1996:9).
99. The drunken devout trope is further highlighted in Lawrence Rosen's (2002:6) *The Culture of Islam*, in which the first chapter features a "thoroughly drunk" Moroccan qadi.
100. Among those who do take note are Asad (1986b:8) and Lukens-Bull (1999:8). Spencer (1989:147) finds the lack of ethnographic documentation a common problem in Geeertz's work.
101. Pals (1996:256), although he later (p. 262) complains that Geertz omits practical information on Islamic worldview.
102. Geertz (1968:vii).
103. Geertz (1968:vii).
104. Sharabi (1990:10).
105. Eickelman (1976:20). Firth (1969:910) found no "systematic consideration of the structure of Islam in the two countries."
106. el-Zein (1977:252).
107. Hourani (1991:100); Martin (1985:10; 1996:40); Smith (1978) and the list goes on virtually *ad infinitum*. Martin (1996:40) adds that the historian of religion adds cosmology and pathos to Geertz's worldview and ethos. It appears that some scholars, e.g., Waldman (1985:95–96), find it sufficient simply to quote and praise Geertz's definition of religion rather than apply it.
108. Launay (1992:31).
109. Geertz (1968:101).

Chapter 2   Ernest Gellner: Idealized to a Fault

1. Gellner (1981:85).
2. Gellner (1981:vii).
3. Asad (1986b:2) observes that the theoretical problems in Gellner's text are emblematic of faults in writing about Islam by other anthropologists, "Orientalists, political scientists, and journalists." Gellner cracked a sharp wit throughout his corpus and the critique of his work has been anything but quiescent, as shall be seen.

4. Geertz (2000:62).
5. Gellner (1981:16).
6. Geertz (1983) and Crapanzano (1986).
7. For critiques of Gellner's segmentary lineage fetish, see Asad (1986a) reprinted in Asad (1993:171–199); Asad (1986b); Anderson (1984); Eickelman (1982); Munson (1993b:278); Roberts (2002:107); and Rosen (2002:41–55). Asad (1979:622) characterizes Gellner's philosophical approach as a "Wizard of Oz theory of ideology." Rosen (2002:41–42) effectively closes the book: "Like the drunk who searches for his keys under the lamppost not because that is where he lost them but because that is where he imagines the most available light to exist, the pursuit of 'the truth' about segmentarity is ripe for disregard, the cutting of one's losses, and (as many long since have) moving on to other things."
8. Quoted in Roberts (2002:107). Even Gellner's most sustained critics, e.g., Henry Munson (1995:831) maintain the highest respect for the scholar and the man. His words and ideas are what need to be assessed. My conscious parody of Gellnerian caustic prose is meant to discredit what he says and not who he was.
9. Gellner (1981:vii).
10. The ellipsis is found in the frontispiece citation.
11. The frontispiece indicates that this comes from *Arabia* [*sic*], but in fact it is from a review in *Arabica*.
12. Virtually no other ethnographic accounts are consulted in *Muslim Society*. Gellner read widely in the ethnographic literature, including unpublished Ph.D. dissertations, but almost none of this appears relevant to making his arguments here.
13. Gellner (1969:3).
14. Gellner (1995:821).
15. Gellner (1981:71).
16. Asad (1986b:8). Kraus (1998:5) notes that Gellner's ethnographic data "do not add up to more than a sketchy account," and as a result Gellner's argument cannot really be empirically refuted. Anderson (1984:117) is similarly harsh in suggesting that "Gellner's adumbration with modern ethnography updates him to the analytical standards of 18th century rationalism."
17. Prakash (1954:494).
18. Al-Azmeh (1984:116).
19. There are a vast number of references, mostly derivative, in Western languages on Ibn Khaldun and his works. By the late 1970s Al Azmeh (1981) uncovered over 850 references on Ibn Khaldun. This is not surprising given that Orientalists have had access to excerpts of his work since Silvestre de Sacy published a few fragments in 1806. A brief account of the author is provided by Talbi (1971). The older work of Schmidt (1930) is still a valuable resource on early citations and commentaries.

20. Fortunately for Ibn Khaldun, he lived more than a century before Ferdinand and Isabella expelled all Muslims and Jews from Spain in 1492.
21. This took place on January 10, 1401. It is reputed that the author let himself down by rope over the wall rather than accepting the offer. See Fischel (1967).
22. Issawi (1950:1). I quote from Issawi's chapter titles, which frame his highly selected excerpts, thus carefully avoiding passages that would link the author more closely to the collective foibles of his age. In a critical review of Issawi's treatment, Cedric Dover (1952:110) complained that the editing looked "like a dull assortment of extracts from textbooks of sociology."
23. Issawi (1950:2).
24. Ibn Khaldun (1958(3):77–78). Lest the reader think I have a personal objection to the scholarship of Ibn Khaldun, I need to make it clear that he is obviously a very important scholar and certainly ahead of his time in the fourteenth century, East or West. The problem, as noted below, is that his ideas have been "modernized" out of context. I do offer, however, a pertinent and ironic Khaldun quote as an epigraph for the perils of contemporary scholarship: "It should be known that among the things that are harmful to the human quest for knowledge and to the attainment of a thorough scholarship are the great number of works (available), the large variety in technical terminology (needed for purposes) of instruction, and the numerous (different) methods (used in those works) . . . His [the student] whole lifetime would not suffice to know all the literature that exists in a single discipline, (even) if he were to devote himself entirely to it . . ." (Ibn Khaldun 1958:3:288).
25. The first European biography of Ibn Khaldun appears in d'Herbelot's *Bibliotheque orientale* in 1697.
26. Ayad (1930:163).
27. Sarton (in Issawi 1950:x).
28. Dover (1952:109).
29. Baali and Wardi (1981:vii); Enan (1941:152); Sarton (in Issawi 1950:x).
30. Sarton (in Issawi 1950:x).
31. Bosch (1950:26).
32. Baali and Wardi (1981:vii); Issawi (1950:17); Prakash (1954).
33. Aziz Al-Azmeh (1984:114–115), who mentions this but does not subscribe to the forced link.
34. Said (1986:150). Said's purpose is largely rhetorical, to explain Foucault's view of history.
35. Baali and Wardi (1981:vii).
36. Issawi (1950:10), who claims that Ibn Khaldun "shows deeper insight than Hobbes."
37. Issawi (1950:14), because he "excludes Philosophy only to make room for Faith."

38. Dover (1952:111).
39. Baali and Wardi (1981:21); Bouthoul (1930); Enan (1961:152), E. Rosenthal (1932); Sarton (in Issawi 1950:x).
40. Baali and Wardi (1981:vii); Joseph Freiherr von Hammer-Purgstall in Enan (1941:150).
41. Bouthoul (1930).
42. Ahmed (1986:217; 1988:101).
43. Baali and Wardi (1981:vii) and Enan (1961:152).
44. Prakash (1954).
45. Myers (1964:60); Prakash (1954); Thomson (1933:110).
46. Issawi (1950:13), in that both seek "neither to praise nor to blame."
47. Myers (1964:58).
48. Lacoste (1984:161).
49. Baali and Wardi (1981:vii); Enan (1961:152); Ghazoul (1992:162); Sarton (in Issawi 1950:x). Sociologist Harry Barnes (1917:198) prefers Ibn Khaldun to Vico as the founder of the philosophy of history.
50. Ahmed (1986:217; 1988:101).
51. Prakash (1954).
52. Davis (1994:189). His concern is understandable, given the kind of statements published in earlier sociological journals; e.g., "Even yet, I would match the seminal ideas of these two ancients [Ibn Khaldun and Machiavelli] against all the products of the sociological research bureaus of our day" (Lee 1955:650). But not all his colleagues heard his message, since Spickard (2001:114) recently writes: "Ibn Khaldun's combination of new concepts with a new framework can help Western sociologists see with a new eye."
53. Except where noted, the quotes in this paragraph are from Issawi (1950:ix–x).
54. Toynbee (3:321–322). For Yves Lacoste (1984:1,6), Ibn Khaldun's work "marks the birth of the science of history," and the conceptions of history by Thucydides, St. Augustine, Machiavelli and Montesquieu are "qualitatively poorer" than Ibn Khaldun's work. A recent uncritical echo of this accumulated acclaim is given by Ghada Osman (2003:50).
55. Issawi (1950:16).
56. Batseva (1971:122, 131). Akbar Ahmed (1986:217; 1988:101) suggests that Ibn Khaldun is "reflected" in Marx.
57. Talbi (1971:830). Gellner (1981:46–47) notes that both Marx and Engels knew of Ibn Khaldun's ideas.
58. Weiss (1995:34).
59. As noted by Lawrence (1984b:12, note 8).
60. Gellner (1981:34–35), who adds "Evidently, Ibn Khaldun held that government should restrict itself to a Keynesian propping up of aggregate demand, but leave the rest to the market, without itself dabbling in enterprise."
61. Gellner (1981:87). Akbar Ahmed (1988:101) trumps Gellner by making the eleventh century al-Biruni "the first major anthropologist

of Islam." Al-Biruni, Ahmed asserts, developed a methodology for the study of caste in India that prefigures the work of Louis Dumont.

62. Gellner (1981:18). My point is that Gellner and numerous others refuse to consider Ibn Khaldun as a wide-ranging scholar in context, but instead reduce him to an icon for a modern idea.

63. Issawi (1950:7–8).

64. Ahmed (1988:101); Dhaouadi (1990); Enan (1962:121). Al-Azmeh (1981:34), by contrast, remarks: "The end result was a lame science of *'umrân* which left its object of investigation undefined . . ." Or, as Lacoste (1984:98) suggests, "Ibn Khaldun is not primarily concerned with studying society in general."

65. Anderson (1984:112–113).

66. Rosenthal in Ibn Khaldun (1958:1,lxvii).

67. Gibb (1962:174). The original article was published in 1933.

68. Gibb (1962:167).

69. Von Grunebaum (1954:339–340, note 39).

70. Dunlop (1971:138); see also Dunlop (1951). Cf. the comments of Hernandez (1994:2:802), "However, Ibn Khaldûn cannot be regarded as the pioneer of positive history, since he did not apply the methodology of the *Muquaddima* to the rest of his work, nor is he the precursor of Hegel, Nietzsche, or Comte, nor the antecedent of historical materialism, as has been asserted. Ibn Khaldûn's empiricism has very concrete limits . . ."

71. Aziz Al-Azmeh (1984: 114–115).

72. Brunschvig (1947(2):391), as translated by Talbi (1971:831). Another example of this de-Islamicizing of Ibn Khaldun can be found in Schmidt (1930:20): "Because he is confident that there is an intelligible sequence, a causal connection, an ascertainable order of development, a course of human events following observable tendencies, in accordance with definite laws, he also believes that in proportion as history becomes what in its nature it is, it will be able to predict the future." Somehow Schmidt missed the underlying *inshallah*; most assuredly he paid no attention to the *bismillah* at the beginning of the text.

73. Lawrence (1984b:5), although it must be remembered that Ibn Khaldunism also appeared among sixteenth century Ottoman scholars (Fleischer 1984).

74. Makdisi (1970), quoted in Sivan (1985:48). Sivan (1985:45–72) provides a dated, at times one-sided, discussion of Arab revisionist historians in the 1960s and 1970s.

75. See Lawrence (1984a:81–82). Taha Hussein (1917) doubts that Ibn Khaldun was even an Arab, challenging his family tree as dubious.

76. For example, an article on the intellectual output of Malik Bennabi designates him as "the most original Arab thinker since Ibn Khaldun when it came to speculating on the phenomenon of civilization" (Fahmi Jad'an 1979:401, quoted in Bariun 1992:327).

77. This is discussed in Ibn Khaldun (1958(2):353ff). Elsewhere Gellner (1969:5) calls this a "theory of tribal circulation of élites."

78. Issawi (1950:10), Mahdi (1957:196). It is well to consider Anderson's (1984:119) warning that *'asabiyya* "is an abstraction, like the sociological concept of social solidarity, which points to rather than accounts for phenomenon [*sic*], and which is not so much explained as serves to explain."

79. Gellner (1969:6).

80. Baali and Wardi (1981:96).

81. Ayad (1930:203).

82. This is how de Slane generally presents the term in his 1862–1868 French translation, but Lacoste (1984:103) recounts other phrasing used in this translation.

83. Gabrieli (1930).

84. Enan (1962:114).

85. Dover (1952:119), who states there is "no book more important to the history of racial ideas" than the *Muqaddima*.

86. Fischel (1967:153).

87. Ritter (1948).

88. Lacoste (1984:100), who discusses some of the major definitions prior to 1966.

89. Gellner (1969:7).

90. Gellner (1969:6). This mischievous canine metaphor has been adopted by Lindholm (1996:21), who provides an equally idiosyncratic account of the history of Islam as a proto-American egalitarian value system. For a critique of Gellner's binary misconstruction of Islam, see Abu-Zahra (1997:37–38).

91. I assume that by "anthropolatry" Gellner means an excessive worship of the human over the divine. The absurdity of this contrast is evident in a passage from Ibn Khaldun, quoted in *Muslim Society* (Gellner 1981:18), in which sedentary people are criticized for being lazy and "sunk in well-being and luxury."

92. Gellner (1969:10–11).

93. Fusfeld (1984:91), for example, criticizes Gellner's failure to examine the ideology of organized sufi orders.

94. Gellner (1981:24), who might have usefully learned from Lacoste (1984:110), originally writing in 1966, that Ibn Khaldun rarely uses the term *'asabiyya* in his discussion of Bedouins.

95. Ibn Khaldun (1958(2):265).

96. Ibn Khaldun (1958(1):168). This crude environmental determinism is hard to root out in the works of those who laud Ibn Khaldun as a scientific writer. Lindholm (1996:40), for example, romantically links the rise of "the Middle Eastern heritage of emissary prophecy," including Islam, to "the struggle for survival in the harshness of the arid environment."

97. Ibn Khaldun (1958(2):266).

98. Gellner (1981:35).

99. The primary discussion about Hume is given by Gellner (1981:7–16) in a section called "David Hume and Islam." All of Gellner's comments cited here are from this section unless otherwise noted.

100. There is a difference between calling Hume's philosophy "sociol-ogy," as Gellner loosely attempts, and regarding the *The Natural History of Religion* as "the first move in what might now be called the sociology of religion" (Gaskin 1978:146).
101. See Hume (1956:21).
102. Hume (1956:23).
103. As quoted by Gaskin (1978:147), Hume believes the following: "The proper office of religion is to reform men's lives, to purify their hearts, to enforce all moral duties, and to secure obedience to the laws and civil magistrate." As such, he is closer to a mason than a sociologist. But Gaskin glosses over the irony in Hume's remarks; Hume saw over-revered monotheism as one of the most corrupting influences on morality. To the extent the devout monotheist claims to believe something that is not based on experience, it is better to be an idolater who is at least reckoning from impressions of the experiential realm (Terry Godlove, p.c.).
104. I assume that replacing "flux and reflux" with "oscillation" is Gellner's unique usage. The term does not appear in either Hume's writings or the sources I have consulted.
105. Gellner (1981:9), who omits a comma in his quote from Hume (1956:46).
106. Hume (1956:46–47). This is the first paragraph of chapter VIII.
107. In the late nineteenth century both Herbert Spencer and Edward Tylor built on this idea to offer a "dream theory" of religion as a psychological response.
108. Hume (1956:48).
109. Hume (1956:65).
110. Hume (1956:73).
111. Evans-Pritchard (1965:103). The issue of original monotheism or polytheism came to the fore after the European discovery of natives in various parts of the "New World." This issue took on a political overtone with theorizing that there had been more than one cre-ation and that since New World peoples were not descended from Adam, they did not have to be treated according to the Ten Commandments. In this debate Hume sided with the polygenists (Harris 1968:87).
112. Needham (1972:12).
113. Gellner (1981:8), quoting Williams (1973).
114. For interpreters who take Hume literally on this point, see Herdt (1997:171) and Root (in Hume 1956:7).
115. Gellner (1981:11). Hume's contradiction, for Gellner, is that he praises classical paganism but deplores Catholicism as a solid Scottish protester of popery. This is only a contradiction if it is assumed that Hume thought "paganism" in any guise a mortal sin; Hume, however, was not writing theology.
116. Gellner (1981:14).
117. At times Gellner (1981:14) finds Hume more convincing than Weber.
118. Hume (1956:45).

119. Hume (1956:50). It is, I suggest, little compensation to Muslims that Hume is less hard on Islam, about whom he was no doubt quite uninformed, than Catholics who burn heretics at the stake. "And the same fires, which were kindled for heretics," predicts Hume (1956:54), "will serve also for the destruction of philosophers."

120. Hume (1956:56). Hume earlier quotes Ibn Rushd, Averroes, that "of all religions, the most absurd and nonsensical is that, whose votaries eat, after having created, their deity."

121. Hume (1962:321). I am indebted to Ira Singer for drawing my attention to Hume's anti-Catholic rhetoric in several of his seminal texts. A major point of the anecdote is to emphasize the rational choice made by Mustapha: a piece of bread cannot in substantial or transsubstantial meaning also, at the same time, be the body of a resurrected god (Terry Godlove, p.c.).

122. Gellner (1981:195). I extract the short sentence for rhetorical play. The context is relevant: ". . . A favourite explanation with journalists commenting on the Moroccan scene: do not attempt to make consistent sense of Moroccan affairs. It *never* makes sense. It is all a matter of personalities, irrational passions, accidents, etc., etc. No explanations are possible or required for the conduct of these people. This is facile and false. But the frequency with which journalists are forced to invoke such pseudo-explanations, when pressed to attempt any kind of coherent account, is significant." Indeed, such nonexplanation is facile and false. But the alternative is not to wrap up a complex issue too neatly in arcane philosophical trappings.

123. Anderson (1984:117).

124. Gellner (1981:1). For an earlier critique of Gellner's phrasing here, see Asad (1986b:3–5).

125. Fascination with de Tocqueville also motivates Lindholm (1996:xiii) in his Gellnerian sequel billed as "historical anthropology."

126. Gellner (1981:7). At this point, it might be said, it was not entirely clear how far Europe had been Christianized.

127. Kharejite here refers to an early group of "secessionists" who held to the egalitarian principle that any decent Muslim could be caliph. They were ruthlessly suppressed and blamed for many of the ills of the known world.

128. Gellner's dialectical relation to Hegel is more than mere imagination; Leela Gandhi (1998) accuses Gellner of having "a Hegelian bias."

129. Gellner (1981:48).

130. Durkheim (1965:59) wrote: "A society whose members are united by the fact that they think in the same way in regard to the sacred world and its relations with the profane world, and by the fact that they translate these common ideas into common practices, is what is called a Church. In all history, we do not find a single religion without a Church." Regardless of Durkheim's ethnocentric choice of terminology, it is obvious that Islam is no exception.

131. Gellner (1981:48). Note how Gellner twists the Durkheimian sense of "church," since Durkheim—rightly or wrongly, it does not matter here—insisted that there was no church of magic.
132. Asad (1986b:3).
133. Gellner (1981:58).
134. Gellner (1981:60). The examples given would indicate that most forms of fascism and autocratic rule could be labeled "Koranic."
135. Gellner (1981:62).
136. Hefner (1997:20).
137. Gellner (1981:39). See Gellner (1969:41–68) for his initial theoretical formulation, which draws heavily on the earlier work of Evans-Pritchard's analysis both of the Sudanese Nuer and Cyrenaican Bedouin. Following the critique of Munson (1993b), Gellner (1995:821) responded that he had mainly concentrated on the saints and not on tribal organization.
138. Gellner (1981:84).
139. Gellner (1981:69).
140. Munson (1993b). See also the critiques by Cornell (1998:106–107); Dresch (1984:45); and Hammoudi (1974). For a bibliographic review of the debate over segmentation, see Eickelman (2002:138). Steven Caton (1987) provides an excellent discussion of the controversy.
141. Asad (1986b:9–11).
142. Gellner (1981:81). In this sentence pastoralism is first claimed to incline imitation of segmentary politics but then the pastoral ethos is said to be far more obliging. So is segmentary organisation inevitable or not?
143. Gellner accepts the mercifully disabused dichotomy of desert vs. sown as paradigmatic for representing Islamic history. "This fusion of scripturalism and pastoralism, the implications of each pushed à *outrance* in one continuous system, *is* the classical world of Islam," argues Gellner (1981:24). Not only is there no "once continuous system" in the history of Islamic societies, but the classical world was anything but Bedouinized from without. Infighting among the powerful and later from wily mercenaries, more than solidarity-solid nomads from the steppe, played havoc with dynasties. I do not think Ibn Khaldun had in mind the Mongol hordes when developing his model of dynastic rise and fall.
144. The book's first chapter is replete with comparison to the history of Christianity to the point that Islamic civilization "seems a kind of mirror-image of traditional Christendom" (Gellner 1981:54).
145. Martin and Woodward (1997:224).
146. Cornell (1998:xxvii). As Anderson (1984:117) avers, Gellner as exegete is a "decontextualizing synthetic philosopher" rather than a historian.
147. This criticism has been made by Asad (1986b:8). Horvatich (1997:184) notes that Gellner's contrast between tribesmen and townsmen treats villages as "closed communities."

148. Ironically, Eickelman (1976) ignored Gellner's (1969) *Saints of the Atlas* in his discussion of previous treatment of the "maraboutic crisis." Gellner (1981:214–220), despite a glowing disclaimer that Eickelman's ethnography is "a very thoroughly researched, sensitively interpreted, elegantly and readable presented case study," proceeds to accuse Eickelman "in common with other members of his particular school" of "exaggerating," "underestimating," "vacillating"—or seeming to—and presenting "a curious mistake." The real target of Gellner is actually Geertz, Eickelman's assumed main mentor. But Eickelman (1982:572) returns the disfavor, referring to Gellner's unique combination of Hume and Ibn Khaldun as *bricolage*. Munson (1993b:269) questions Gellner's understanding of his informant's words and complains that ethnographic evidence is not provided to support the claim for a segmentary lineage model among the Ait 'Atta.

149. Gellner (1969:xiii). The two specific time periods mentioned in the 1950s are summer vacations.

150. Gellner (1981:70).

151. Peters (1960, 1967).

152. I do not include in this listing those figures mentioned in quotes or elliptically: they are—in alphabetical order—Abraham, al-'Abbas, [St.] Anthony, [St.] Benedict, Brasidas, Byron, Caleb, [St.] Dominic, [St. Francis], Hector, Hercules, Isaac, Jacob, Joshua, Kierkegaard, Mahdi of Khartoum, Muhammad ibn 'Abd al-Wahhab, Muhammad ibn Sa'ud, Romulus, Theseus and the Virgin Mary. The interested reader can uncover yet more names in the 99—a highly significant number for any study of Islam—footnotes.

153. Gellner's precise reference is to "anti-Milton-Friedman pleas."

154. Gellner quotes from their *Marxisme et Algérie* (1976), a title which would seem to qualify them as sociological.

155. If you are curious why Ibn Khaldun is placed here, I simply follow Gellner's insistence that he is "the greatest sociologist of Islam." I also take the liberty of cross-referencing Ibn Khaldun as an Arab.

156. Gellner first identifies him as a "Russian scholar" and later as a "Soviet anthropologist" and "Soviet authority on the Scythians."

157. Authors of *L'Algérie des Anthropologues* (1975).

158. Described by Gellner as the "self-appointed Grand Mufti of the anthropologists." Unless Mead had an unnoticed sex-change, this clever nomen would seem at cross purposes.

159. I assume that a "Soviet specialist on nomadism" fits here as well as anywhere.

160. This is the only female "authoress" quoted in the essay, although Gellner does not find her argument plausible. He also mentions the work of his student, Shelagh Weir.

161. Gellner muses about a film spectacular financed by British Petroleum and starring Omar Sharif as this Methodist.

162. A translator of Ibn Khaldun.

163. Gellner feels compelled to point out that he is "a convinced Christian and Quaker" with a "rather Augustinian/Kierkegaardian picture."
164. As in "the Wittfogelian sense."
165. The reference is actually to General Daumas, the French consul to the amir.
166. Grandson of the prophet, murdered by Yazid.
167. Also referred to in a revolutionary way as "Che Khomeiny."
168. The fictional sociologist who might have authored *The Kharejite Ethic and the Spirit of Capitalism*, if Poitiers had been a Western defeat.
169. Gellner (1989:10) prefaces his later *Plough, Sword and Book* with a tradition of Muhammad.
170. A fictional philosopher. Gellner might have noted that such a savant would no doubt have left us pondering the bulbul of Baghdad rather than the owl of Minerva. In a later passage Gellner (1981:115) finds the latter a "much overrated bird."
171. The "murderer of Hussein," who was a son of 'Ali.
172. This is my own interpolation; I am not aware where Gellner ended up on the intellectual search for a historical Jesus.
173. The English biographer of Amir Abd el Kader, not the British politician who said "Some chicken. Some neck."
174. To be fair, the reference is to "Napoleon's army."
175. Mentioned in order to paraphrase him.
176. Gellner (1981:2, 11, 27, 54, 71, 79).
177. As Eickelman (1982:572) retorts, "the alternatives to a procrustean, all-encompasssing model for 'Islamic Society' are sociological explanations animated by an attention to specific Islamic societies and their concrete historical developments, including those of the long middle periods." Rosen (2002:52) similarly complains that in fact alternative theories had been presented.

Chapter 3   *Beyond the Veil:* At Play in the Bed of the Prophet

1. Mernissi (1987:45).
2. The popular American stereotype is well summed up by Bruce Lawrence (1998:5): "Behind the hostile Muslim men, Americans imagine the faces of Muslim women, homebound creatures marked alike by seclusion from the outside world and apparent oppression by their tyrannical husbands. The reality of Muslim women's active participation in their societies is glossed, covered, as it were, by a veil that projects the violence of male 'Arab' Muslims everywhere. They hate the West and abuse their women."
3. A major bibliographic resource to start with is Kimball and von Schlegell (1997). Tapper and Tapper (1987:88) have argued that the elucidation of gender and religious orthodoxy "must be central to any anthropology of Islam." Saba Mahmoud (2001a) cautions

against *a priorizing* a Western liberal feminist voice as the default anthropological position.

4. Muir (1912:333).
5. Said (1979:12).
6. Said (1994:xxv).
7. For an expansion on Abbott's pioneering study, see Spellberg (1994).
8. Muir (1912:290).
9. Abbott (1942:vii).
10. Muir (1912:xlviii).
11. Since the onset of the internet, the chances of a student taking the time to shake the dust off such a volume have decreased. However, given the unevenness of information readily available at the fingertip on the web, the problem of outmoded expertise remains.
12. Muir (1912:200).
13. Muir (1912:23).
14. Muir (1912:333ff).
15. Muir (1912:335–336).
16. Mabro (1991:197–222).
17. This is in reference to surah (33:51); see Muir (1912:295).
18. Muir (1912:300). For a discussion of the Western gaze at the imagined siren of the seraglio, see Schick (1999) and Yeazell (2000). Mabro (1991) provides relevant excerpts from a number of Western travelers, male and female.
19. Abbott (1942:vii).
20. Abbott (1942:ix).
21. Abbott (1942:viii).
22. Abbott (1942:7).
23. Abbott (1942:22).
24. Abbott (1942:66).
25. Abbott (1942:23).
26. The Mormons had often been criticized as an apostasy like Islam; e.g., Kinney (1912).
27. The phrase is from Said (1979:328).
28. A. Rahman (1986:2, 10).
29. Bint al-Shati (1971:12). For a summary of this commentator's views on women in Islam, see Hoffmann-Ladd (1987:36–38).
30. For information on Muslim veneration of Muhammad, see Schimmel (1985).
31. See Stowasser (1994:85–103).
32. Haykel (1976:xxxv).
33. Haykel (1976:li). On Haykel's work, Rodinson (1981:29) comments: "It is a skillful reconstruction of the life of Muhammad suited to the needs of a modern apologetic, but it is far from being scientific in its viewpoint."
34. Haykel (1976:lxxv).
35. Haykel (1976:lxiii).
36. Haykel (1976:lx).

37. Haykel (1976:63).
38. Haykel (1976:289).
39. Haykel (1976:292).
40. Bint al-Shati' (1971:12). For a mystical interpretation of the value of Muhammad's physical love for his wives, see Ibn al-Arabi's thirteenth century commentary in Kvam et al. (1999:200–203).
41. Bint al-Shati' (1971:14).
42. Tradition quoted by Mernissi (1987:43).
43. Combs-Schilling (1989:58).
44. Mabro (1991).
45. Only the Dutch title, *Achter de Sluier*, maintains the same metaphor. The French, Spanish, and German versions are variations of "Sex, Ideology, Islam". The Arabic, *Al-Jins ka-handasa ijtima'iya*, basically means "Sex as Social Construction." More details on Mernissi's books are available at her website, <www.mernissi.net>.
46. Martin and Woodward (1997:206).
47. Bullock (2002:136).
48. Mernissi (1987:xvi). Martin and Woodward (1997:207) suggest that Mernissi was influenced by Foucault and Said in her second edition. Given the fact that neither scholar is quoted directly, nor any sense of discourse is articulated, this appears to be an example of reading theory into a text that is virtually uninformed by contemporary critical theory.
49. Mernissi (1987:27–28).
50. Malti-Douglas (1991:43–44) criticizes Mernissi for relying solely on "programmatic manuals" like that of al-Ghazali. Bullock (2002:163–169) argues that Mernissi misreads al-Ghazali. Al-Ghazali's text contains many contradictory statements and traditions which other scholars have interpreted differently.
51. Mernissi (1989:31).
52. See Kassis (1983:449). See Gardet (1965) for a succinct discussion of the term's usage.
53. Surah 64:15. There are examples in Egyptian Arabic in which Muslim women refer to horrible husbands as producing *fitna* (Mahmoud (2001a:219).
54. Mernissi (1987:54), who simply cites al-Bukhari as her source. In a later work, Mernissi (1991:49–61) questions the authenticity of this specific tradition; my concern here is only with the impact of *Beyond the Veil*.
55. El Guindi (1999:25).
56. Mernissi (1987:46). In a polemical work published in India, Mernissi (1986) calls on Muslim women to form an "International Association of Women Interested in Designing Alternative Paradises."
57. Mernissi (1987:84–85).
58. Suad Joseph (1977:468) concludes: "Mernissi's new and interesting perspective, however, is compromised by its theoretical confusion, undisciplined methodology, and episodic style." Nikkie Keddie

(1979:227) refers to Mernissi's modeling of Muslim gender as "entirely ideal." See the critiques by Bullock (2002:139) on Mernissi's ahistorical and reductive approach, El Guindi (1999:25) on Mernissi's use of "Christano-European feminist ideology," Majid (2000:107–109) on Mernissi's uncompromising secular approach and Tapper (1979).

59. Eickelman (1989:202, note 60); this observation has been dropped in the 4th edition. Janet Abu-Lughod (1977:365) also forgave Mernissi the many flaws in her book because the author held "deep conviction."

60. Mernissi (1987:8).

61. Mernissi (1987:89–90). For this book Mernissi only conducted formal interviews and examined an Islamic variant of "Dear Abby" letters; she conducted no fieldwork.

62. Mernissi (1987:ix).

63. As Jonah Blank (2001:123) observes, "One of the most common misperceptions about Islam prevalent in non-Islamic circles is that the religion is fundamentally antagonistic to the freedom and high status of women." *Beyond the Veil* does nothing to discourage such misperception. Unfortunately, Mernissi is not alone in essentializing a Muslim concept of gender. Anthropologist Charles Lindholm (1995:815) refers to the egregious assumptions in Mernissi's work, but in his [a]historical [mis]anthropology of the Islamic Middle East he succumbs to generic nonsense such as the following: "Given these negative factors, it is no surprise that everywhere in the Middle East, when a boy baby is born, it is the occasion of noisy congratulations, while silence or condolences greet the birth of a girl" (Lindholm 1996:230). When did "everywhere" become an ethnographic fieldsite?

64. Mernissi (1987:52).

65. Mernissi (1987:53). Martin and Woodward (1997:209) are critical of Mernissi's rereading of the early intellectual history of Islam.

66. Mernissi (1987:53).

67. Combs-Schilling (1989:92–93).

68. Combs-Schilling (1989:93, 95).

69. Combs-Schilling (1989:95).

70. Combs-Schilling (1989:97).

71. Combs-Schilling (1989:97), who cites this tradition to justify her claim that Muslims are supposed to enjoy sex with a woman but not be "psychologically attached" to her. She is apparently unaware that al-Ghazali includes this tradition in a section devoted to "comfort and relaxation for the soul through companionship" (al-Ghazali in Farah 1984:1965). Compare the reading of Ibn al-'Arabi that this saying shows that male perfection lies in women (Murata 1992: 183, 186–195).

72. Combs-Schilling (1989:68, 69). For a critique of the Western feminist essentialization of Muslim women, see Mohanty (1994).

73. Combs-Schilling (1989:61).

74. Combs-Schilling (1989:242).
75. Rosen (2002:179, note 40).
76. Bowen (1992:658).
77. Delaney (1998:181).
78. Boudhiba (1985:12).
79. Combs-Schilling (1989:54). Similarly, Delaney (1985:283) relies on established "orientalist" renderings of Islam by Gibb and Lewis. Inexplicably, Edward Said is not mentioned in Combs-Schilling's book, but William Safire is.
80. Delaney (1991) also quotes widely from Boudhiba. In a footnote (1991:320, note 37) she rightly criticizes his lurid portrayal of the afterlife as an "infinite orgasm" for being a very sexist view. Yet, Boudhiba's (1985:172) symbolic equation of the *hamam* and female anatomy is accepted by Delaney as an obvious truth.
81. Combs-Schilling (1989:70).
82. Anderson (1982:398).
83. For a comprehensive survey of how the "veil" has been represented, especially through ethnography, see El Guindi (1999).
84. Haykel (1976:285).
85. Daniel (1960:98). In his pseudo attack on Muhammad as the icon of imposture, Humphrey Prideaux (1697:143) writes that "the old lecher fell desperately in love" with Zaynab. Ironically, St. Thomas Aquinas recognized that Muslims made a counter claim of sexual dalliance: "Truly the Saracens deride us because we say Christ is the Son of God, when God had no wife . . ." (Quoted in Waltz 1976:85). Djait (1985:14) speculates that the obsession of celibate medieval monks about Muhammad's sex life may have stemmed in part from their own abhorrence of carnal pleasure.
86. Muir (1912:292).
87. Muir (1912:294, note 1).
88. Abbott (1942:18).
89. Andrae's *Mohammed, Sein Leben und Sein Glaube* was originally published in 1932 and was eventually translated into Spanish, Italian, and English.
90. Abbott (1942:51).
91. Abbott (1942:21).
92. Abbott (1942:26).
93. Haykel (1976:286).
94. Haykel (1976:288).
95. Bint al-Shati' (1971:147).
96. Bint al-Shati' (1971:144).
97. Bint al-Shati' (1971:153).
98. Mernissi (1987:56–57).
99. Mernissi (1987:42–43). The danger in fixing on a specific tradition out of context is that other traditions can be found with a very different angle. Muhammad is also quoted as saying: "Paradise lies beneath the feet of the mothers" (quoted in Schimmel 1985:51).

Consider the irony in the following tradition quoted by Schimmel (1985:46): "One day a little old woman came to him to ask whether old wretched women would also go to Paradise. 'No,' answered the prophet, 'there are no old women in Paradise!' And then, looking at her grieved face, he continued with a smile: 'They will all be transformed in Paradise for there, there is only one youthful age for all!' "

100. Al-Ghazali (in Farah 1984:62) notes that Muhammad once found himself attracted to a woman and warded off temptation by returning home to have sex with his wife Zaynab.

101. Mernissi (1991:86–87).

102. Mernissi (1987:95). For an extended critique of Mernissi's reading of the Zaynab passage, see Bullock (2002:177). Leila Hessini (1994:54) suggests that the veil serves a paradoxical purrpose because it can also serve as "a liberating force from some Moroccan women."

103. Mernissi (1991:10).

104. Muir (1912:522).

105. Mernissi (1987:8).

106. Consider the comments of Katherine Bullock (2002:180): "In ignoring covered women's voices and in reducing them to passive victims, Mernissi is only reinscribing the colonial and Orientalist view of the 'veiled woman.' "

107. Abbott (1942:19).

108. Bint al-Shati' (1971:14).

109. Mernissi (1987:33–34).

110. Mernissi (1987:33). *Beyond the Veil* avoids comments by Muslim intellectuals who have criticized denigration of women. For example, Ibn al-'Arabi, an early thirteenth-century Sufi scholar, asserts that "everything to which a man can attain—stations, levels, or attributes—can also belong to any woman whom God wills, just as it can belong to any man whom God wills" (quoted in Murata 1992:183).

111. Quoted in Kvam et al. (1999:132). It is worth noting that Tertullian made this comment in a homiletic on women's proper apparel. Leila Ahmed (1992:68) argues that the views of al-Ghazali were closer to those of Tertullian and St. Augustine than what is articulated in the Quran.

112. Muir (1912:294, note 1).

113. Mernissi (1987:42–43).

114. Mernissi (1987:59).

115. Mernissi (1991:15).

116. Mernissi (1991:17).

117. A similar criticism has been leveled by Bullock (2002:139), who suggests Mernissi adopts an "ahistorical approach to the meanings of religious symbols that fails to contextualize how people enact Islam differently in different times and places."

118. Mernissi (1991:vi).

119. Bullock (2002:178).
120. For a nuanced study of gender roles in the highland valley of al-Ahjur, see Adra (1998).
121. Haeri (1989).
122. Anderson (1982:418).
123. el-Zein (1977:252).
124. Mernissi (1987:xvi; 1991:10).
125. Mernissi (1994).
126. Mohammed Fadel (1997:185–186) thinks that Mernissi's search for an essential Islamic message of gender equality "verges on the absurd" because there are "as many teachings on gender as the different social circumstances in which it is interpreted."
127. Mernissi (1991:11). Amına Wadud (1999:ix) also posits the equality of men and women based on a "female inclusive reading" of the Quran.
128. Abbott (1942:ix).
129. See Cooke (1999:98–101).
130. For example, Mahmoud (2001a); Minh-ha (1989); Strathern (1987); and several of the articles in Behar and Gordon (1995).
131. Abu-Lughod (1990, 1993); El Guindi (1999).
132. Tucker (1993:viii). In this regard, it is important to note the pioneering work of Sachiko Murata (1992) in using the Asian model of *yin* and *yang* to understand gender in Islam.
133. Tapper (1979:481).
134. Murata (1992:4). This might best be labeled the white woman's burden.
135. As an example of how our own culture-bound views about sexuality influence our understanding of Islamic views, consider Delaney's (1991:51) summary account of her discussion with Turkish informants about oral sex: "Given the extreme emphasis on the penis, I found it hard to believe that fellatio was not more common, yet the response by this group of women leads me to believe that it is rare; still, I hesitate to generalize about such an intimate subject." What are the objective criteria for asserting that her Turkish villagers probably have an "extreme emphasis on the penis?" Why was it so hard to take these women's responses at face value? I find it ironic that Delaney does not recognize such an extreme emphasis on the divine phallus inserted by Combs-Schilling into her interpretation of the Abrahamic sacrifice motif.

Chapter 4   Akbar Ahmed: Discovering Islam Inside Out

1. Ahmed (1988:8).
2. Evans-Pritchard (1956:322).
3. Ahmed (1988:5). Ironically, being South Asian is so important to Ahmed that he refers to Salman Rushdie as "the only Muslim to win

the Booker Prize." Rushdie's *The Satanic Verses* appeared the same year as Ahmed's *Discovering Islam*. See Ahmed (1992:163–177) for his later take on Rushdie.

4. Young (1996:131–139).
5. It is important to make a distinction between a secular approach which does not rule out the role of faith and an apologetic religious stand. Omid Safi (2003:20) argues for a progressive Islam which moves beyond "apologetic presentations of Islam."
6. Ahmed (1988:3). I choose this as a representative book of Ahmed's extensive written corpus, one which is somewhat involuted by publishing virtually identical essays in different forums (see the editor's note in Ahmed 1983:81). This book was reprinted in 2002 with a new introduction. In addition, Ahmed was the advisor and star of a 1993 BBC series called "Living Islam: What it Means to be a Muslim in Today's World," which is reviewed by Titus (1995).
7. Taken from the back cover promotion of the original paperback edition of *Discovering Islam* (Ahmed 1988).
8. Ahmed (1988:ix).
9. In a later text, Ahmed (1999:234) makes it clear that Khomeini's brand of Islam "sees the West as the primary enemy both culturally and politically." Earlier in this text Ahmed (1999:110) provides a confused interpretation of Khomeini's views on Islam. He seems to accept at face value that Khomeini saw all Islam as one rather than simply *shi'a*, when in fact that oneness was thoroughly *shi'a*; the ideal is misread as the real.
10. Ahmed (1988:ix–x).
11. I refer here to inconsistencies in Ahmed's argument as well as a journalistic writing style that begs for a better editor. Critical reviews of the style and content of *Discovering Islam* include Donnan (1989) and Metcalf (1989). Titus (1995:543) notes Ahmed's "tendency to slip into overly simple generalizations."
12. Ahmed (1976, 1980).
13. The author quotes extensively, but fails to provide a bibliography of references cited. In the "suggested readings," there is not one work of a Muslim historian, not even Ibn Khaldun. The only anthropologist mentioned besides Ahmed and his collaborator for an edited volume on *Islam in Tribal Societies* is Gellner, whose *Muslim Society* is recommended reading.
14. Ahmed (1988:1).
15. Ahmed (1988:2).
16. Ahmed (1988:2).
17. Schlegel (1998). This intriguing ethnography, subtitled "The Spiritual Journey of an Anthropologist" is a semi-autobiographical account by a man who started out as an Episcopalian missionary, but later returned to study an isolated forest community as an anthropologist.
18. For example, Ahmed (1988:8), where he suggests that "the role of the neutral social scientist is almost mythical."

19. Ahmed (1988:9). His tone is often apologetic, even if not overtly polemical, as noted by Jon Anderson (1981) in a review of Ahmed (1980).

20. Ahmed (1988:9). Critics of Ahmed's work note that he has a tendency, as Dale Eickelman (in Ahmed 1983:83) suggests, to assume no one else is doing what he does and to either ignore or misread relevant scholarly research within and outside anthropology; see also Tapper (1995:193, note 7).

21. Ahmed (1988:11).

22. The significant event before this is the assassination of Anwar Sadat in 1981; before that in 1979 we learn that Khomeini overthrows the shah, Soviet troops enter Afghanistan and Juhaiman attempts to seize the Ka'ba.

23. Weber (1964:1). This was originally written in 1922, before any of the modern ethnographic corpus on Muslim social behavior was available.

24. Ahmed (1988:3).

25. Ahmed (1988:4).

26. el-Zein (1977).

27. Ahmed (1988:3).

28. Ahmed (1988:31). I assume the "we" here refers to fellow Muslims who agree with the author. The choice of "flux and reflux" here is probably inspired by Gellner (1981).

29. Ahmed (1988:55).

30. Ahmed (1988:31).

31. Ahmed (1988:32). Ahmed (1988:109) discusses this *jihad* as an example of Islam confronting challenges by inspiring men to "action and piety." He does not mention that it also pitted Muslim against Muslim at the time over interpretation of syncretic elements in local Islam among the Hausa.

32. Ahmed (1988:35).

33. Ahmed (1988:35).

34. Ahmed (1988:49).

35. Ahmed (1988:50).

36. Ahmed (1988:51).

37. Ahmed (1988:35–36). "The saints and scholars of Shiaism are revolutionaries, exhorting, pointing towards the ideal," adds Ahmed (1988:59). So does this put the blame for missing the ideal on the *sunni* majority?

38. Ahmed (1988:59).

39. Ahmed (1988:212).

40. Ahmed (1986:217 and 1989:234). Ahmed (1988:212–216) also provides a shorter account of Islamic anthropology in *Discovering Islam*. Similar calls for a distinctive Islamic anthropology have been made by Davies (1988) and Maroof (1981), as well as the earlier work of Ali Shariati (1979). Tapper (1995) provides a review of these efforts.

41. Ahmed (1988:215).

42. Ahmed (1988:213).

43. Ahmed (1984:3).

44. Ahmed (1988:214).
45. Ahmed (1988:214). An earlier variant is provided in Ahmed (1984:3) in which *jihad* is glossed as "to better the world."
46. Said (1979:328).
47. Ahmed (1988:99). In an earlier assessment, Ahmed (1984:3) dubbed this scholar the "first professional and practising anthropologist."
48. Al-Biruni (1973:8).
49. See al-Biruni (1879:90), who was defending miraculous aspects in sacred history from what many today would regard as rational criticism.
50. Ali in Ahmed (1983:81). Ali also criticizes Ahmed for idealizing and thus misrepresenting the roles of mullah and pir in Pakistan.
51. Ahmed (1983:85). Could we say that the U.S. nuclear program is Christian because it evolved to meet the threat of the godless Russian communists?
52. The quoted phrase is from Ahmed (1988:214).
53. See Ahmed (1988:215). It hardly suffices to answer fellow Muslims who see an unbridgeable gap between Islam and the West by citing well-known Arab scholars of centuries past as proto-anthropologists.
54. The proceedings were published in 1989. The paper delivered by Ahmed (1989) was published earlier (Ahmed 1986) and outlined in Ahmed (1984).
55. This is taken from the Introduction to Abu Saud et al. (1989); no author is indicated for the introduction.
56. Muhammad (1989:230).
57. Ma'ruf (1989:166). While he draws attention to evolution as a key issue for developing an Islamic view of anthropology, the article is a descriptive history, now woefully out of date, with no suggestions on how to reconcile modern evolutionary theory with Islamic theology.
58. Tapper (1995:191), who fears that those proposing the idea of an Islamic anthropology are not open to constructive criticism.
59. In *Postmodernism and Islam* Ahmed (1992) cites the canon of post-modern criticism with a vengeance, pun intended, but only to indict it as the wellspring of modern secular materialism.
60. The phrase is from Ahmed (1986:193).
61. Ahmed (1986:226).
62. See Varisco (2002a:63–64) for a discussion of nineteenth-century Christian interpretation of bible history through Bedouin eyes.
63. Ahmed (1986:181).
64. Tapper (1995:193).
65. Ahmed (1988:215). Gellner, it would seem, gets it right in Ahmed's eyes because he borrows his theory from Ibn Khaldun (Ahmed 1988:101).
66. Abdul-Rauf (1985:187).
67. Turner (1994:12), who does not regard Ahmed's (1992) *Postmodernism and Islam* as up to the task.
68. Ahmed (1992). I take these from the book's index, but citations are strewn throughout the narrative.

Epilogue   Muslims Observed: The Lessons
From Anthropology

1. Arkoun (1994:2).
2. Arkoun (1994:16) provides a similar parody of book titles in which Christianity is applied as indiscriminately as Islam. Ironically, Joel Robbins (2003) has recently called for an anthropology of Christianity given the "success" of the anthropology of Islam. However, Robbins' goal is not to invent a palatable concept for a Western audience, as has been the case in representing Islam, but to exorcise the spectre of the Christian heritage haunting Euro-American anthropologists.
3. I follow the example of Talal Asad (1986b:7), who in challenging the essentialized notions of Islam by Gellner and Geertz, asks "What kinds of questions do these styles deflect us from considering? What concepts do we need to develop as anthropologists in order to pursue those very different kinds of questions in a viable manner?"
4. Jackson (1989:2). This is the anthropologist, not the singer.
5. Digard (1978:497).
6. Gellner taught most of his life in a sociology department. Gilsenen (2000:5) and Tapper (1984) routinely refer to the "sociology of Islam" as synonymous with anthropology. In a recent compendium entitled *Islam: Critical Concepts in Sociology*, Turner (2003) includes a selection from Gellner, as well as several excerpts from American anthropologists.
7. Eickelman (1976:4). Anthropologists Fischer and Abedi (1990:xxi) likewise refer to the "sociological" texture of the Islamic culture of Iran.
8. Hefner (2000:xi), who defines his own text as both historical sociology and social anthropology.
9. Ethnographers at times are present when government census surveys are taken; see Bradburd (1998:9–19) for an example from Iran.
10. Asad (1986b:12).
11. Delaney (1991:25).
12. Martin (1996:244).
13. Thomson (1901:31–32), where the author is described as "thirty years missionary in Syria and Palestine."
14. Bradburd (1998:57).
15. Holy (1991:219).
16. Holy (1991:6).
17. Bowen (1997:159).
18. Consider, in this respect, the anecdote used by David Hume (1956:56) about Mustapha's spin on the trinity, or, to probe more deeply into the intellectual baggage of the Western study of religion, the canard about the Bororo parrot (Smith 1978).
19. Bradburd (1998:161).
20. Gellner (1969:305), who makes light of his unfamiliarity with both Berber and Arabic, at one point stating he stands by the sound of a

name he heard, even though it is a phonetic impossibility (p. 306). No phonetic transcriptions are provided in his text, although he discusses several local legends.

21. Gellner (1969:303).
22. Notable exceptions are the honest appraisals by Daniel Bradburd (1998), mentioned above, and Bill Young (1996).
23. Bradburd (1998:61).
24. Bradburd (1998:132).
25. Young (1996:21).
26. el-Zein (1977:252).
27. Safi (2003:20).
28. Asad (1986b:5).
29. el-Zein (1977:227).
30. el-Zein (1977:251).
31. Recently, several scholars have returned to el-Zein's point about multiple "islams" (Hussein 2003:268, note 14; Kassam 2003:142, note 3).
32. Asad (1986b:2), who, along with Eickelman (2002:245), is interested in the institutional and discursive structures that produce social knowledge. For a nuanced critique of Asad, see Lukens-Bull (1999).
33. Asad (1986b:11).
34. El Guindi (1999:67) and Delaney (1991:19). Although el-Zein's (1974) ethnography of Muslims on Lamu does not pursue a functionalist goal, as defined by Durkheim or Malinowski, his sophisticated analysis of the symbols associated with local readings of "Islam" is hardly a denial of the meaning interpretable out of commonly shared sources.
35. Launay (1992:5).
36. Robbins (2003:194).
37. el-Zein (1977:252). This view of culture is not strictly Lévi-Straussian, as Asad implies, but parallels the debate in linguistics at that time about the nature of language.
38. el-Zein (1974:xx).
39. el-Zein (1977:252). John Bowen (2003:20) makes a similar argument: "Nowhere is there an 'Islamic society,' if that phrase implies people simply applying a single set of texts to social life; everywhere there is one if that phrase implies people struggling to rethink those texts in the light of alternative cultural and legal norms."
40. Omid Safi (2003:18).
41. For example, Lawrence (1998:4). Gilsenen (2000:4) notes that in writing his original *Recognizing Islam*, he "did not consider Islam to be a monolithic 'it,' an entity which could be treated as a theological or civilisational historical bloc, unchanging and essentially 'other' in some primordial ways."
42. Chick (1988:20). This comic is part 6 in a series about an alleged former Catholic priest named Alberto Rivera, who is born again into Chick's brand of Christianity. The series sets out to prove that the world can be divided into born-again Christians versus all the rest. The rest, including Catholics, Communists and Muslims, are all

under the power of Satan, God's cosmic adversary. Details on Chick and his conversion-oriented tracts and comics can be found at his main website <www.chick.com>.

43. Morey (1992), whose subtitle says it all: "Confronting the World's Fastest Growing Religion." Morey's elementary-school style of writing and total disdain for scholarly dialogue will not get it on a college reading list, but it is likely to sell well at Walmart. Esposito's (1992) influential book has gone through three editions and has recently been updated in his *Unholy War: Terror in the Name of Islam* (2002). For a similarly well-crafted critique of those who perpetuate negative stereotypes, see Lawrence (1998).

44. Lewis (2001), who apparently likes the sound of the phrase, since he has recycled it in articles and book sections over many years.

45. Lewis (2001:51). At the risk of being facetious, I wonder if it is not Lewis who has failed to appreciate history by speculating before all the data are in. In criticism of American spokesmen who he said at the time refused to implicate Iraqi involvement in the 9/11 attack, Lewis (2001:63) surmised that Saddam Hussein would indeed be a "formidable adversary" who would "not be restrained by any scruples" in using his "considerable arsenal of unconventional weapons."

46. Edwards (2002).

47. Hefner (1997:14).

48. Lukens-Bull (1999:10).

49. el-Zein (1977:227).

50. Regarding her fieldwork on Tunisian rain rituals, Nadia Abu-Zahra (1997:4) concludes. "The fieldwork data would have been incomprehensible had I not consulted the Qur'an and the Arabic works of the commentators on the prophet's traditions."

51. Asad (1986b:14).

52. Abu-Zahra (1997); Antoun (1989); el-Aswad (2000); el-Zein (1974).

53. Launay (1992); Messick (1993).

54. Horvatich (1994:820).

55. Delaney (1991:30).

56. Abu-Zahra (1997:4).

57. El Guindi (1999:xiii).

58. Peters (1984:214). In the same volume Richard Tapper (1984:249) states that "the stereotype of the impious nomad, like that of the nomad implacably hostile to settled life, has no general validity."

59. Abu-Zahra (1997:83–286).

60. Bowen (1993:7).

61. Lambek (1993:66).

62. El Guindi (1999:xv).

63. Messick (1993:1).

64. Dresch and Haykel (1995).

65. A similarly perceptive parsing of politics and religion is provided in Jenny White's (2002) ethnographic analysis of Islamist mobilization in secular Turkey.

66. See Varisco (1995) for details.
67. For example, 'Adnan's son Ma'add was said to be a contemporary of the Babylonian king Nebuchadnezzar (al-Tabari 1988:37).
68. For a specific ethnographic example of this point, see Peters (1960).
69. Lambek (1993:11).
70. Delaney (1991:19).
71. I say this not to avoid criticism, but to acknowledge that I personally cannot be an ethnographer without realizing what it means to be male, relatively well off economically, forged in my formative years by a distinctive Christian upbringing and, like it or not, an American in a part of the world where America is more often reviled as the Great Satan rather than a promised land with streets paved of gold. For an example of how two anthropologists of Islamic Iran define postmodernity, see Fischer and Abedi (1990:xxxi–xxxii).
72. El Guindi (1999:xv). This is not just a recent issue. As Horvatich (1994:816) notes, America's imperial designs in the Philippines included attempts to discourage Islam and Christianize the population.
73. This catastrophe stimulated a barrage of media commentary and paperback bestsellers. Out of the mix I recommend the informative and readable study by John Esposito (2002).
74. See especially Shahrani (2002). One senior anthropologis, Ashraf Ghani, accepted a cabinet post in the post-Taliban Afghan government.
75. Armbrust (2002).
76. Hefner (2002).
77. Schneider and Schneider (2002).
78. Shryock (2002).
79. Bowen (1996:12).
80. I borrow this apt metaphor from Jack Renard.
81. Huntington (1993).
82. Hefner (2000); Blank (2001).
83. Lukens-Bull (1999:1).
84. Asad (1986b:17).
85. Bourdieu (1990).
86. Delaney (1991:21). See Goodman (2003); Mahmoud (2001b); and Starrett (1995) for similar critiques of Bourdieu. In fairness to Bourdieu, his earlier published work on Algeria indicates a concern in demystifying the notion of Islam as a dogma that causes cultural phenomena (Bourdieu 1962:108).
87. Asad (2003:17).
88. Digard (1978:498).
89. The idea of culture is so seductive that the latter part of the last century witnessed the proliferation of a field called Cultural Studies.

Bibliography

Abaza, Mona and Georg Stauth 1988. Occidental Reason, Orientalism, Islamic Fundamentalism: A Critique. *International Sociology* 3(4): 343–364.

Abbott, Nabia 1942. *Aishah: The Beloved of Muhammad*. Chicago: University of Chicago Press.

Abdul-Rauf, Muhammad 1985. Outsiders' Interpretations of Islam: A Muslim's Point of View. In Richard C. Martin, editor, *Approaches to Islam in Religious Studies*, 175–188. Tucson: University of Arizona Press.

Abu-Lughod, Janet 1977. *Beyond the Veil: Male-Female Dynamics in a Modern Muslim Society*, by Fatima Mernissi . . . *Contemporary Sociology* 6: 364–366 (Review).

Abu-Lughod, Lila 1990a. Anthropology's Orient: The Boundaries of Theory on the Arab World. In Hisham Sharabi, editor, *Theory, Politics and the Arab World: Critical Responses*, 81–131. London: Routledge.

——— 1990b. Can There be a Feminist Ethnography? *Women & Performance* 5(1): 7–27.

——— 1993. Finding a Place for Islam: Egyptian Television Serials and the National Interest. *Public Culture* 5: 493–513.

Abu Saud, Mahmoud et al. 1989. *Toward Islamization of Disciplines*. Herndon, VA: International Institute of Islamic Thought.

Abu-Zahra, Nadia 1970. On the Modesty of Women in Arab Muslim Villages: A Reply. *American Anthropologist* 72: 1079–1087.

——— 1997. *The Pure and the Powerful: Studies in Contemporary Muslim Society*. Reading: Ithaca Press.

Adams, Charles J. 1985. Foreword. In Richard C. Martin, editor, *Approaches to Islam in Religious Studies*, vii–x. Tucson: University of Arizona Press.

Adra, Najwa 1997. The "Other" as Viewer: Reception of Western and Arab Televised Representations in Rural Yemen. In Peter I. Crawford and Sigurjon B. Hafsteinsson, editors, *The Construction of the Viewer*. Proceedings from NAFA 3, 255–269. Denmark: Intervention Press.

——— 1998. Dance and Glance: Visualizing Tribal Identity in Highland Yemen. *Visual Anthropology* 11: 55–102.

Ahmed, Akbar S. 1976. *Millennium and Charisma among Pathans: A Critical Essay in Social Anthropology*. London: Routledge, Kegan Paul.

——— 1980. *Pukhtun Economy and Society: Traditional Structure and Economic Development in a Tribal Society*. Boston: Routledge, Kegan Paul.

Ahmed, Akbar S. 1983. Islam and the District Paradigm. *Current Anthropology* 24(1): 81–87.

——— 1984. Defining Islamic Anthropology. *RAIN* #65: 2–4.

——— 1986. Toward Islamic Anthropology. *The American Journal of Islamic Social Sciences* 3(2): 181–230.

——— 1988. *Discovering Islam: Making Sense of Muslim History and Society*. London: Routledge.

——— 1989. Toward Islamic Anthropology. In Mahmoud Abu Saud et al., *Toward Islamization of Disciplines*, 199–247. Herndon, VA: International Institute of Islamic Thought.

——— 1992. *Post Modernism and Islam: Predicament and Promise*. London: Routledge.

——— 1999. *Islam Today: A Short Introduction to the Muslim World*. London: I. B. Taurus.

Ahmed, Akbar S. and David M. Hart, editors 1984. *Islam in Tribal Societies: From the Atlas to the Indus*. London: Routledge and Kegan Paul.

Ahmed, Leila 1992. *Women and Gender in Islam*. New Haven: Yale University Press.

al-Birini 1879. *The Chronology of Ancient Nations*. Translated by C. Edward Sachau. London: William H. Allen and Co.

——— 1973. *Al-Biruni's Book on Pharmacy and Materia Medica*. Edited and translated by Hakim Mohammed Said. Karachi: Hamdard National Foundation.

al-Hibri, Azizah 1982. A Study of Islamic Herstory: or How Did We Ever Get into this Mess? In Azizah al-Hibri, editor, *Women and Islam*, 207–219. Oxford: Pergamon Press.

al-Tabari 1987ff. *The History of Al-Tabari*. Albany: SUNY Press.

Altorki, Soraya 1986. *Women in Saudi Arabia: Ideology and Behavior among the Elite*. New York: Columbia University Press.

Altorki, S. and C. F. El-Solh, editors 1988. *Arab Women in the Field: Studying Your Own Society*. Syracuse: Syracuse University Press.

Ammar, Hamed 1954. *Growing Up in an Egyptian Village*. New York: Grove Press.

Anderson, Jon W. 1981. Pukhtun Economy and Society: Traditional Structure and Economic Development in a Tribal Society (Review). *American Ethnologist* 8(2): 400–402.

——— 1982. Social Structure and the Veil: Comportment and the Composition of Interaction in Afghanistan. *Anthropos* 77: 397–420.

——— 1984. Conjuring with Ibn Khaldun: From an Anthropological Point of View. In Bruce B. Lawrence, editor, *Ibn Khaldun and Islamic Ideology*, 111–121. Leiden: Brill.

——— 1998. The Internet and Islam's New Interpreters. In Dale F. Eickelman and Jon W. Anderson, editors, *New Media in the Muslim World: The Emerging Public Sphere*, 41–56. Bloomington: Indiana University Press.

Andrae, Tor 1932. *Mohammed, Sein Leben und Sein Glaube*. Göttingen: Vandenhoeck and Ruprecht.

Andriolo, Karin 2002. Murder by Suicide: Episodes from Muslim History. *American Anthropologist* 104(3): 736–742.

Antoun, Richard 1968a. On the Modesty of Women in Arab Muslim Villages: A Study in the Accommodation of Traditions. *American Anthropologist* 70: 671–697.

——— 1968b. The Social Significance of Ramadan in an Arab Village. *The Muslim World* 58: 36–42, 95–104.

——— 1972. *Arab Village: A Social Structural Study of a Transjordanian Peasant Community*. Bloomington: Indiana University Press.

——— 1976. Anthropology. In Leonard Binder, editor, *The Study of the Middle East*, 137–213. New York: John Wiley and Sons.

——— 1980. The Islamic Court, the Islamic Judge and the Accommodation of Traditions: A Jordanian Case Study. *International Journal of Middle East Studies* 12: 455–467.

——— 1989. *Muslim Preacher in the Modern World: A Jordanian Case Study in Comparative Perspective*. Princeton: Princeton University Press.

Arkoun, Mohammed 1994. *Rethinking Islam: Common Questions, Uncommon Answers*. Translated by Robert D. Lee. Boulder: Westview Press.

Armbrust, Walter 1996. *Mass Culture and Modernism in Egypt*. Cambridge: Cambridge University Press.

——— 2000. The Riddle of Ramadan: Media, Consumer Culture, and the Christmas-ization of a Muslim Holiday. *Working Papers of the MES*. Electronic Document, http://www.aaanet.org/mes/arm.htm. Accessed April, 2004.

——— 2002. Islamists in Egyptian Cinema. *American Anthropologist* 104(3): 922–931.

Asad, Talal 1979. Anthropology and the Analysis of Ideology. *Man*, (n.s.) 14: 607–627.

——— 1980. Ideology, Class and the Origins of the Islamic State. *Economy and Society* 9: 450–473.

——— 1983. Anthropological Conceptions of Religion: Reflections on Geertz. *Man* (n.s.) 18: 237–259.

——— 1986a. The Concept of Cultural Translation in British Social Anthropology. In James Clifford and George E. Marcus, editors, *Writing Culture: The Poetics and Politics of Ethnography*, 141–164. Berkeley: University of California Press.

——— 1986b. *The Idea of an Anthropology of Islam*. Occasional Papers Series. Washington, D.C.: Georgetown University Center for Contemporary Arab Studies (reprinted 1996).

——— 1993. *Genealogies of Religion: Discipline and Reasons of Power in Christianity and Islam*. Baltimore: Johns Hopkins Press.

——— 2003. *Formations of the Secular: Christianity, Islam, Modernity*. Stanford: Stanford University Press.

Aswad, Barbara 1970. Social and Ecological Aspects in the Formation of Islam. In Louise Sweet, editor, *Peoples and Cultures of the Middle East*, 1: 53–73. Garden City: Natural History Press.

Ayad, M. Kamal 1930. *Die Geschichts-und Gesellschaftslehre Ibn Haldûns.* Stuttgart: J. G. Cotta'sche.

Al-Azmeh, Aziz 1981. *Ibn Khaldun in Modern Scholarship: A Study in Orientalism.* London: Third World Centre for Research and Publishing.

———— 1984. The Articulation of Orientalism. In Afaf Hussain et al., editors, *Orientalism, Islam, and Islamists,* 89–124. Brattleboro: Amana Books.

Baali, Fuad and Ali Wardi 1981. *Ibn Khaldun and Islamic Thought-Styles: A Social Perspective.* Boston: G. K. Hall.

Bahloul, Joëlle 1996. *The Architecture of Memory: A Jewish-Muslim Household in Colonial Algeria 1937–1962.* Cambridge: Cambridge University Press.

Banks, David J. 1990. Resurgent Islam and Malay Rural Culture: Malay Novelists and the Invention of Culture. *American Ethnologist* 17(3): 531–548.

Barclay, Harold B. 1963. Muslim Religious Practice in a Village Suburb of Khartoum. *Muslim World* 53: 205–211.

———— 1964. *Buuri al Lamaab: A Suburban Village in the Sudan.* Ithaca: Cornell University Press.

Bariun, Fawzia 1992. Malik Bennabi and the Intellectual Problems of the Muslim Ummah. *The American Journal of Islamic Social Sciences* 9(3): 325–337.

Barnes, Harry E. 1917. Sociology before Comte: A Summary of Doctrines and Introduction to the Literature. *American Journal of Sociology* 23(2): 174–247.

Barrett, Stanley A. 1996. *Anthropology: A Student's Guide to Theory and Method.* Toronto: University of Toronto Press.

Barth, Fredrik 1964. *Nomads of South Persia: The Basseri Tribe of the Khamseh Confederacy.* Oslo: Universiteitsforlaget.

Bates, Daniel G. 1973. *Nomads and Farmers: A Study of the Yörük of Southeastern Turkey.* Museum of Anthropology Anthropological Papers, 52. Ann Arbor: University of Michigan.

Bates, Daniel G. and Amal Rassam 2001. *Peoples and Cultures of the Middle East.* Second Edition. Upper Saddle River: Prentice-Hall.

Batseva, S. M. 1971. The Social Foundations of Ibn Khaldûn's Historico-Philosophical Doctrine. *Islamic Quarterly* 15: 121–132.

Bauerlein, Mark 1997. *Literary Criticism: An Autopsy.* Philadelphia: University of Pennsylvania Press.

Behar, Ruth and Deborah A. Gordon, editors 1995. *Women Writing Culture.* Berkeley: University of California Press.

Benda, Harry J. 1962. *The Religion of Java . . . Journal of Asian Studies* 21: 403–406 (Review).

Benedict, Ruth 1934. *Patterns of Culture.* Boston: Houghton Mifflin.

Biddick, Kathleen 1994. Bede's Blush: Postcards from Bali, Bombay, Palo Alto. In John Van Engen, editor, *The Past and Future of Medieval Studies,* 16–44. Notre Dame: University of Notre Dame.

Binder, Leonard 1988. *Islamic Liberalism: A Critique of Development Theories.* Chicago: University of Chicago Press.

Bint al-Shati' 1971 [1959]. *The Wives of the Prophet*. Translated by M. Moosa and D. N. Ranson. Lahore: Sh. Muhammad Ashraf.

Blank, Jonah 2001. *Mullahs on the Mainframe: Islam and Modernity among the Daudi Bohras*. Chicago: University of Chicago Press.

Bloch, Marc 1953. *The Historian's Craft*. New York: Vintage Books.

Bodman, Herbert L. 1991. *Women in the Muslim World*. Providence: AMEWS.

Boon, James 1982. *Other Tribes, Other Scribes: Symbolic Anthropology in the Comparative Study of Cultures, Histories, Religions, and Texts*. Cambridge: Cambridge University Press.

Bosch, Gular Kheirallah 1950. Ibn Khaldun on Evolution. *The Islamic Review* 38: 26.

Boudhiba, Abdelwahab 1985 [1970]. *Sexuality in Islam*. London: Routledge and Kegan Paul.

Boulakia, J. D. C. 1971. Ibn Khaldûn: A Fourteenth Century Economist, *Journal of Political Economy* 79: 1105–1118.

Bourdieu, Pierre 1960. *The Algerians*. Translated by Alan Ross. Boston: Beacon Press.

———— 1990 [1980]. *The Logic of Practice*. Stanford: Stanford University Press.

Bouthoul, G. 1930. *Ibn Khaldun, sa philosophie sociale*. Paris: Paul Geuthner.

Bowen, John R. 1989. *Salât* in Indonesia: the Social Meanings of an Islamic Ritual. *Man* 24(4): 600–619.

———— 1992. On Scriptural Essentialism and Ritual Variation: Muslim Sacrifice in Sumatra and Morocco. *American Ethnologist* 14(4): 656–671.

———— 1993. *Muslims through Discourse: Religion and Ritual in Gayo Society*. Princeton: Princeton University Press.

———— 1996. Religion in the Proper Sense of the Word: Law and Civil Society in Islamicist Discourse. *Anthropology Today* 12(4): 12–14.

———— 1997. Modern Intentions: Reshaping Subjectivities in an Indonesian Muslim Society. In Robert W. Hefner and Patricia Horvatich, editors, *Islam in an Era of Nation-States: Politics and Religious Renewal in Muslim Southeast Asia*, 157–181. Honolulu: University of Hawaii Press.

———— 2003. *Islam, Law, and Equality in Indonesia: An Anthropology of Public Reasoning*. Cambridge: Cambridge University Press.

Bradburd, Daniel 1998. *Being There: The Necessity of Fieldwork*. Washington D.C.: Smithsonian Institution Press.

Brantlinger, Patrick 1990. *Rule of Darkness: British Literature and Imperialism, 1830–1914*. Ithaca: Cornell University Press.

Brunschvig, Robert 1947. *La Berbérie orientale sous les Hafsides des origines à la fin du XV siècle*. Paris: Adrien-Maisoneuve.

Bullock, Katherine 2002. *Rethinking Muslim Women and the Veil: Challenging Historical & Modern Stereotypes*. Herndon: The International Institute of Islamic Thought.

Burton, Richard F. 1855–56. *Personal Narrative of a Pilgrimage to Al-Madinah and Meccah*. 3 volumes. London: Longmans.

Canfield, Robert Leroy 1973. *Faction and Conversion in a Plural Society: Religious Alignments in the Hindu Kush*. Museum of Anthropology Anthropological Papers, 50. Ann Arbor: University of Michigan.

Caton, Steven C. 1987. Power, Persuasion, and Languages: A Critique of the Segmentary Model in the Middle East. *International Journal of Middle East Studies* 19(1): 77–101.

Chelhod, Joseph 1958. *Introduction à la sociologie de l'islam: de l'animisme à l'universalisme.* Paris: G.-P. Maisonneuve.

———— 1969. Ethnologie du monde Arabe et Islamologie. *L'Homme* 9(4): 24–40.

Chick, Jack 1988. *The Prophet. Alberto Series Part Six.* Chino, CA: Chick Publications.

Clifford, James 1988. *The Predicament of Culture: Twentieth-Century Ethnography, Literature, and Art.* Cambridge: Harvard University Press.

Clifford, James and George E. Marcus, editors 1986. *Writing Culture: The Poetics and Politics of Ethnography.* Berkeley: University of California Press.

Cole, Donald 1975. *Nomads of the Nomads: The Al Murrah Bedouin of the Empty Quarter.* Chicago: Aldine.

Combs-Schilling, Elaine 1989. *Sacred Performances: Islam, Sexuality, and Sacrifice.* New York: Columbia University Press.

Cooke, Miriam 1999. Feminist Transgressions in the Postcolonial World. *Critique* 14: 93–106.

Coon, Carlton S. 1951. *Caravan, the Story of the Middle East.* New York: Henry Holt.

Cornell, Vincent 1998. *Realm of the Saint: Power and Authority in Moroccan Sufism.* Austin: University of Texas Press.

Crapanzano, Vincent 1973. *The Hamadsha: A Study in Moroccan Ethnopsychiatry.* Berkeley: University of California Press.

———— 1986. Hermes' Dilemma: The Masking of Subversion in Ethnographic Description. In James Clifford and George E. Marcus, editors, *Writing Culture: The Poetics and Politics of Ethnography,* 51–76. Berkeley: University of California Press.

D'Andrade, Roy 1995. *The Development of Cognitive Anthropology.* Cambridge: Cambridge University Press.

Daniel, Norman 1960. *Islam and the West: The Making of an Image.* Edinburgh: Edinburgh University Press.

Davies, Wyn 1988. *Knowing One Another: Shaping Islamic Anthropology.* London: Mansell.

Davis, James A. 1994. What's Wrong with Sociology? *Sociological Forum* 9(2): 179–197.

Davis, John 1991. An Interview with Ernest Gellner. *Current Anthropology* 32: 63–72.

Dawud, N. J., translator 1968. *The Koran.* Baltimore: Penguin Books.

Delaney, Carol 1990. The *Hajj:* Sacred and Secular. *American Ethnologist* 17(3): 513–530.

———— 1991. *The Seed and the Soil: Gender and Cosmology in Turkish Village Society.* Berkeley: University of California Press.

———— 1998. *Abraham on Trial: The Social Legacy of Biblical Myth.* Princeton: Princeton University Press.

Dhaouadi, Mahmoud 1990. Ibn Khaldûn: The Founding Father of Eastern Sociology. *International Sociology* 5(3): 319–335.

Digard, Jean-Pierre 1978. Perspectives anthropologiques sur l'islam. *Revue francaise de Sociologie* 19: 497–523.

Djait, Hichem 1985. *Europe and Islam.* Translated by Peter Heinegg. Berkeley: University of California Press (original, 1978).

Donnan, Hastings 1989. *Discovering Islam: Making Sense of Muslim History and Society.* Akbar S. Ahmed . . . *Man*, N. S., 24: 350–351 (Review).

Donner, Fred M. 1981. *The Early Islamic Conquests.* Princeton: Princeton University Press.

Douglas, Mary 1966. *Purity and Danger.* London: Routledge and Kegan Paul.

Dover, Cedric 1952. The Racial Philosophy of Ibn Khaldun. *Phylon* 13(2): 107–119.

Dresch, Paul 1984. The Position of Shaykhs among the Northern Tribes of Yemen. *Man* 19: 31–49.

Dresch, Paul and Bernard Haykel 1995. Stereotypes and Political Styles: Islamists and Tribesfolk in Yemen. *International Journal of Middle East Studies* 27(4): 405–431.

Dunlop, Douglas Morton 1951. An Arab Philosophy of History . . . *Philosophical Quarterly* 1: 473–474 (Review).

————— 1971. *Arab Civilization to A.D. 1500.* London: Longman.

Dupree, Louis 1973. *Afghanistan.* Princeton: Princeton University Press.

Durkheim, Emile 1965. *The Elementary Forms of the Religious Life.* New York: The Free Press (original, 1912).

Duvignaud, Jean 1970. *Change at Shebika: Report from a North African Village.* New York: Pantheon Books.

Edwards, David B. 1995. Print Islam: Media and Religious Revolution in Afghanistan. *Anthropological Quarterly* 68(3): 171–184.

————— 1996. *Heroes of the Age: Moral Fault Lines on the Afghan Frontier.* Berkeley: University of California Press.

————— 2002. *Before Taliban: Genealogies of the Afghan Jihad.* Berkeley: University of California Press.

Ehrenfels, Omar Rolf 1940. Ethnology and Islamic Studies. *Islamic Culture* 14: 434–446.

Eickelman, Dale F. 1967. Musaylima: An Approach to the Social Anthropology of Seventh Century Arabia. *Journal of the Economic and Social History of the Orient* 10: 17–52.

————— 1976. *Moroccan Islam: Tradition and Society in a Pilgrimage Center.* Austin: University of Texas Press.

————— 1982. Muslim Society. *Man*, (n.s.) 17: 571–572 (Review).

————— 1985. *Knowledge and Power in Morocco.* Princeton: Princeton University Press.

————— 1989. *The Middle East: An Anthropological Approach.* 2nd ed. Englewood Cliffs: Prentice-Hall.

————— 2002. *The Middle East and Central Asia: An Anthropological Approach.* 4th ed. Englewood Cliffs: Prentice-Hall.

Eickelman, Dale F. and Jon W. Anderson 1997. Print, Islam, and the Prospects for Civic Pluralism: New Religious Writings and Their Audiences. *Journal of Islamic Studies* 8(1): 43–62.

——— 1999. Redefining Muslim Publics. In Dale F. Eickelman and Jon W. Anderson, editors, *New Media in the Muslim World: The Emerging Public Sphere*, 1–18. Bloomington: Indiana University Press.

Eickelman, Dale F. and James Piscatori, editors 1990. *Muslim Travellers: Pilgrimage, Migration, and the Religious Imagination*. Berkeley: University of California Press.

Eickelman, Dale F. and Jon W. Anderson, editors 1999. *New Media in the Muslim World: The Emerging Public Sphere*. Bloomington: Indiana University Press.

el-Aswad, el-Sayed 2000. Muslim Sermons: Public Discourse in Local and Global Scenario. *Working Papers of the MES*. Electronic Document, http://www.aaanet.org/mes/aswad.htm. Accessed April, 2004.

——— 2002. *Religion and Folk Cosmology: Scenarios of the Visible and Invisible in Rural Egypt*. Westport: Praeger.

El Guindi, Fadwa 1981. The Emerging Islamic Order: The Case of Egypt's Contemporary Islamic Movement. *Journal of Arab Affairs* 1: 245–261.

——— 1999. *Veil: Modesty, Privacy and Resistance*. Oxford: Berg.

Ellen, Roy F. 1983. Social Theory, Ethnography, and the Understanding of Practical Islam in South-East Asia. In M. B. Hooker, editor, *Islam in South-East Asia*, 50–91. Leiden: Brill.

el-Zein, Abdul Hamid 1974. *The Sacred Meadows*. Evanston: Northwestern University Press.

——— 1977. Beyond Ideology and Theology: The Search for the Anthropology of Islam, *Annual Review of Anthropology* 6: 227–254.

Enan, Mohammed Abdullah 1941. *Ibn Khaldun: His Life and Work*. Lahore: Sh. Muhammad Ashraf.

Esack, Farid 1999. *On Being a Muslim*. Oxford: Oneworld.

Esposito, John L. 1992. *The Islamic Threat: Myth of Reality?* Oxford: Oxford University Press.

——— 2002. *Unholy War: Terror in the Name of Islam*. Oxford: Oxford University Press.

Evans-Pritchard, E. E. 1949. *The Sanusi of Cyrenaica*. Oxford: Clarendon Press.

——— 1956. *Nuer Religion*. Oxford: Oxford University Press.

——— 1966. *Theories of Primitive Religion*. Oxford: Oxford University Press.

Fadel, Mohammed 1997. Two Women, One Man: Knowledge, Power and Gender in Medieval Sunni Legal Thought. *International Journal of Middle East Studies* 29: 185–204.

Farah, Madelaine, translator 1984. *Marriage and Sexuality in Islam*. Salt Lake City: University of Utah Press.

Fernea, Robert and J. M. Malarkey 1975. Anthropology of the Middle East and North Africa: A Critical Assessment. *Annual Review of Anthropology* 4: 183–206.

Firth, Raymond 1969. *Islam Observed . . . Journal of Asian Studies* 28: 909–910 (Review).
——— 1981. Spiritual Aroma: Religion and Politics. *American Anthropologist* 83(3): 582–601.
Fischel, Walter J. 1967. *Ibn Khaldun in Egypt.* Berkeley: University of California Press.
Fischer, Michael 1980. *Iran: From Religious Dispute to Revolution.* Cambridge: Harvard University Press.
Fischer, Michael and M. Abdeli 1990. *Debating Muslims: Cultural Dialogues in Tradition and Postmodernity.* Madison: University of Wisconsin Press.
Fleischer, Cornell 1984. Royal Authority, Dynastic Cyclism, and 'Ibn Khaldûnism' in Sixteenth-Century Ottoman Letters. In Bruce B. Lawrence, editor, *Ibn Khaldun and Islamic Ideology*, 46–68. Leiden: Brill.
Fusfeld, Warren 1984. Naqshabandi Sufism and Reformist Islam. In Bruce B. Lawrence, editor, *Ibn Khaldun and Islamic Ideology*, 89–110. Leiden: Brill.
Gabrieli, Francesco 1930. Il concetto della 'asabiyya nel pensiero storico di Ibn Haldun. *Atti della Reala Accademica della Scienze di Torino*, 65.
Gandhi, Leela 1998. *Postcolonial Theory: A Critical Introduction.* New York: Columbia University Press.
Gardet, Louis 1965. FITNA. In *The Encyclopaedia of Islam* (New Edition), 2: 930–931. Leiden: Brill.
Gaskin, John C. A. 1978. *Hume's Philosophy of Religion.* London: The MacMillan Press.
Geertz, Clifford 1960. *The Religion of Java.* New York: Free Press.
——— 1966. Religion as a Cultural System. In Michael Banton, editor, *Anthropological Approaches to the Study of Religion*, 1–46. London: Tavistock.
——— 1968. *Islam Observed.* Chicago: University of Chicago Press.
——— 1973. *The Interpretation of Cultures.* New York: Basic Books.
——— 1979. Suq: The Bazaar Economy in Sefrou. In C. Geertz, H. Geertz and L. Rosen, editors, *Meaning and Order in Moroccan Society: Three Essays in Cultural Analysis*, 123–313. Cambridge: Cambridge University Press.
——— 1982. Conjuring with Islam. *The New York Review of Books* 29(9): 25–28 (May 27).
——— 1988. *Works and Lives: The Anthropologist as Author.* Stanford: Stanford University Press.
——— 2000. *Available Light: Anthropological Reflections on Philosophical Topics.* Princeton: Princeton University Press.
Gellner, Ernest 1969. *Saints of the Atlas.* Chicago: The University of Chicago Press.
——— 1973. Post-Traditional Forms in Islam: The Turf and Trade, and Votes and Peanuts. *Daedalus* (Winter): 191–206.
——— 1981. *Muslim Society.* Cambridge: Cambridge University Press.
——— 1989. *Plough, Sword and Book: The Structure of Human History.* Chicago: University of Chicago Press.

Gellner, Ernest 1995. Segmentation: Reality or Myth? *The Journal of the Royal Anthropological Institute* 1(4): 829–832.

Gerholm, Tomas and Yngve Georg Lithman, editors 1988. *The New Islamic Presence in Western Europe.* London: Mansell.

Gibb, H. A. R. 1949. *Mohammedanism.* Oxford: Home University Library.

—— 1962. *Studies on the Civilization of Islam.* London: Routledge & Kegan Paul.

Gilsenen, Michael 1973. *Saint and Sufi in Modern Egypt: An Essay in the Sociology of Religion.* Oxford: Oxford University Press.

—— 1982. *Recognizing Islam: An Anthropologist's Introduction.* London: Croom Helm.

—— 1990. Very Like a Camel: The Appearance of an Anthropologist's Middle East. In Richard Fardon, editor, *Localizing Strategies,* 222–239. Washington: Smithsonian Institution Press.

Gladney, Dru C. 1998. *Ethnic Identity in China: The Making of a Muslim Minority.* Case Studies in Social Anthropology. New York: Harcourt Brace Publishers.

—— 1999. The Salafiyya Movement in Northwest China: Islamic Fundamentalism among the Muslin Chinese. In Leif Manger, editor, *Muslim Diversity: Local Islam in Global Contexts,* 102–149. Nordic Institute of Asian Studies, No 26. Surrey: Curzon Press.

Goodman, Jane E. 2003. The Proverbial Bourdieu: Habitus and the Politics of Representation in the Ethnography of Kabylia. *American Anthropologist* 105(4): 782–793.

Goody, Jack 1961. Religion and Ritual: The Definition Problem. *British Journal of Psychology* 12: 143–164.

Gordon, Joel 1998. Becoming the Image: Words of Gold, Talk Television, and Ramadan Nights on the Little Screeen. *Visual Anthropology* 10: 247–264.

Graham, William A. 1993. Traditionalism in Islam: An Essay in Interpretation. *Journal of Interdisciplinary History* 23(3): 495–522.

Greenblatt, Stephen J. 1994. The Eating of the Soul. *Representations* 48: 97–116.

Gupta, Akhil and James Ferguson 1997. Discipline and Practice: "The Field" as Site, Method, and Location in Anthropology. In A. Gupta and J. Ferguson, editors, *Anthropological Locations: Boundaries and Grounds of a Field Science,* 1–46. Berkeley: University of California Press.

Guthrie, Stewart 1980. A Cognitive Theory of Religion. *Current Anthropology* 21(2): 181–203.

Haeri, Shahla 1989. *Law of Desire: Temporary Marriage in Shi'i Iran.* Syracuse: Syracuse University Press.

Hammoudi, Abdellah 1974. Segmentarité, stratification sociale, pouvoir politique et sainteté: Réflexions sur les thèses de Gellner. *Hespéris* 15: 147–180.

—— 1993. *The Victim and its Masks: An Essay on Sacrifice and Masquerade in the Maghreb.* Chicago: University of Chicago Press.

Handleman, Don 1994. Critiques of Anthropology: Literary Turns, Slippery Bends. *Poetics Today* 15(3): 341–381.

Handler, Richard 1991. An Interview with Clifford Geertz. *Current Anthropology* 32: 603–613.

Harris, Marvin 1968. *The Rise of Anthropological Theory*. New York: Thomas Y. Crowell.

Haykel, Muhammad 1976 [1935]. *The Life of Muhammad*. Plainfield, IN: American Trust Publications.

Hefner, Robert W. 1985. *Hindu Javanese: Tengger Tradition and Islam*. Princeton: Princeton University Press.

——— 1996. Islamizing Capitalism: On the Founding of Indonesia's First Islamic Bank. In Mark R. Woodward, editor, *Toward a New Paradigm: Recent Developments in Indonesian Islamic Thought*, 291–322. Tempe: Arizona State University, Program for Southeast Asian Studies.

——— 1997. Islam in an Era of Nation-States: Politics and Religious Renewal in Muslim Southeast Asia. In Robert W. Hefner and Patricia Horvatich, editors, *Islam in an Era of Nation-States: Politics and Religious Renewal in Muslim Southeast Asia*, 3–40. Honolulu: University of Hawaii Press.

——— 2000. *Civil Islam: Muslims and Democratization in Indonesia*. Princeton: Princeton University Press.

——— 2002. Global Violence and Indonesian Muslim Politics. *American Anthropologist* 104(3): 754–765.

Hefner, Robert W. and Allan Hoben 1991. The Integrative Revolution Revisited. *World Development* 19(1): 17–30.

Herdt, Jennifer A. 1997. *Religion and Faction in Hume's Moral Philosophy*. Cambridge: Cambridge University Press.

Hernandez, Miguel Cruz 1994. Islamic Thought in the Iberian Peninsula. In Salma K. Jayyusi, editor, *The Legacy of Muslim Spain*, 2: 777–803. Leiden: Brill.

Hessini, Leila 1994. Wearing the Hijab in Contemporary Morocco: Choice and Identity. In Fatma Müge Göçek and Shiva Balaghi, editors, *Reconstructing Gender in the Middle East*, 40–56. New York: Columbia University Press.

Hicks, David, editor 1999. *Ritual & Belief: Readings in the Anthropology of Islam*. Boston: McGraw Hill.

Hirschkind, Charles 2001. Civic Virtues and Religious Reason: An Islamic Counterpublic. *Cultural Anthropology* 16(1): 3–34.

Hodgson, Marshall 1974. *The Venture of Islam*. 3 volumes. Chicago: University of Chicago Press.

Hoffman-Ladd, Valerie J. 1987. Polemics on the Modesty and Segregation of Women in Contemporary Egypt. *International Journal of Middle East Studies* 19: 23–50.

Holy, Ladislav 1988. Gender and Ritual in an Islamic Society: The Berti of Darfur. *Man* 23(3): 469–487.

——— 1991. *Religion and Custom in a Muslim Society: The Berti of Sudan*. Cambridge: Cambridge University Press.

Horton, Robin 1960. A Definition of Religion, and Its Uses. *Journal of the Royal Anthropological Institute* 90: 201–226.

Horton, Robin 1993. *Patterns of Thought in Africa and the West.* Cambridge: Cambridge University Press.

Horvatich, Patricia 1994. Ways of Knowing Islam. *American Ethnologist* 21: 811–826.

——— 1997. The Ahmadiyya Movement in Simunul: Islamic Reform in One Remote and Unlikely Place. In Robert W. Hefner and Patricia Horvatich, editors, *Islam in an Era of Nation-States: Politics and Religious Renewal in Muslim Southeast Asia,* 183–206. Honolulu: University of Hawaii Press.

Hourani, Albert 1991. *Islam in European Thought.* Cambridge: Cambridge University Press.

Houston, Christopher 2001. The Brewing of Islamist Modernity: Tea Gardens and Public Space in Istanbul. *Theory, Culture & Society* 18(6): 77–97.

Hume, David 1956. *The Natural History of Religion.* Edited by H. E. Root. Stanford: Stanford University Press (original, 1757).

——— 1962. *A Treatise on Human Nature.* Edited by D. G. C. Macnabb. Cleveland: The World Publishing Company (original, 1739).

Huntington, Samuel P. 1993. The Clash of Civilizations? *Foreign Affairs* 72(3): 22–49.

Hussein, Amir 2003. Muslims, Pluralism, and Interfaith Dialogue. In Omid Safi, editor, *Progressive Muslims on Justice, Gender and Pluralism,* 241–269. Oxford: Oneworld.

Hussein, Taha 1917. *Étude analytique et critique de la philosophie sociale d'Ibn Khaldoun.* Paris: A. Pedone.

Ibn Khaldun 1958. *Ibn Khaldûn. The Muqaddimah.* Translated by Franz Rosenthal. 3 volumes. New York: Pantheon Books.

Inhorn, Marcia C. 1994. *Quest for Conception: Gender, Infertility, and Egyptian Medical Traditions.* Philadelphia: University of Pennsylvania Press.

Issawi, Charles, translator 1950. *An Arab Philosophy of History.* London: John Murray.

Jackson, Michael 1989. *Paths Toward a Clearing.* Bloomington: Indiana University Press.

Janowitz, M. 1963. Anthropology and the Social Sciences. *Current Anthropology* 4: 139, 146–154.

Johns, Anthony 1975. Islam in Southeast Asia: Reflections and New Directions. *Indonesia* 19: 33–55.

Joseph, Suad 1977. Beyond the Veil. *American Anthropologist* 79(2): 468 (Review).

Kassam, Tazim R. 2003. On Being a Scholar of Islam: Risks and Responsibilities. In Omid Safi, editor, *Progressive Muslims on Justice, Gender and Pluralism,* 128–144. Oxford: Oneworld.

Kassis, Hanna E. 1983. *A Concordance of the Qur'an.* Berkeley: University of California Press.

Keddie, Nikki 1979. Problems in the Study of Middle Eastern Women. *International Journal of Middle East Studies* 10(2): 255–240.

Keesing, Roger 1987. Anthropology as Interpretive Quest. *Current Anthropology* 28: 161–176.

Kennedy, John 1967. Mushahara: A Nubian Concept of Supernatural Danger and the Theory of Taboo. *American Anthropologist* 69: 685–702.

Kimball, Michelle R. and Barbara R. von Schlegell, editors 1997. *Muslim Women Throughout the World: A Bibliography*. Boulder: Lynne Rienner Publishers.

Klass, Morton 1995. *Ordered Universes: Approaches to the Anthropology of Religion*. Boulder: Westview Press.

Kramer, Martin 2001. *Ivory Towers on Sand: The Failure of Middle Eastern Studies in America*. Washington, D.C.: The Washington Institute for Near East Policy.

Kraus, Wolfgang 1998. Contestable Identities: Tribal Structures in the Moroccan High Atlas. *Journal of the Royal Anthropological Institute* 4(1): 1–22.

Kroeber, Alfred L. and Clyde Kluckhohn 1952. *Culture: A Critical Review of Concepts and Definitions*. Papers of the Peabody Museum, 47, 1. Cambridge: Peabody Museum.

———— 1963. *Culture: A Critical Review of Concepts and Definitions*. New York: Vintage Books.

Kuper, Adam 1999. *Culture: The Anthropologists' Account*. Cambridge: Harvard University Press.

Kvam, Kristen E. et al., editors 1999. *Eve & Adam: Jewish, Christian, and Muslim Readings on Genesis and Gender*. Bloomington: Indiana University press.

Lagrace, R. O. 1957. The Formation of the Moslem State. *Anthropology Tomorrow* 6(1): 141–155.

Laitin, David D. 1978. Religion, Political Culture, and the Weberian Tradition. *World Politics* 30: 563–592.

Lambek, Michael 1993. *Knowledge and Practice in Mayotte: Local Discourses of Islam, Sorcery, and Spirit Possession*. Toronto: University of Toronto Press.

Lane, Edward 1973. *An Account of the Manners and Customs of the Modern Egyptians*. New York: Dover (Facsimile of the 1860 edition).

Laroui, Abdallah 1976. *The Crisis of the Arab Intellectual: Traditionalism or Historicism?* Translated by Diarmid Cammell. Berkeley: University of California Press.

Launay, Robert 1992. *Beyond the Stream: Islam and Society in a West African Town*. Berkeley: University of California Press.

Lawrence, Bruce B. 1984a. Ibn Khaldun & Islamic Reform. In Bruce B. Lawrence, editor, *Ibn Khaldun and Islamic Ideology*, 69–88. Leiden: Brill.

———— 1984b. Introduction: Ibn Khaldun and Islamic Ideology. In Bruce B. Lawrence, editor, *Ibn Khaldun and Islamic Ideology*, 2–13. Leiden: Brill.

———— 1998. *Shattering the Myth: Islam Beyond Violence*. Princeton: Princeton University Press.

Lee, Alfred McClung 1955. The Clinical Study of Sociology. *American Sociological Review* 20(6): 648–653.

Lewis, Bernard 2001. The Revolt of Islam. *The New Yorker* 77(36): 50–63, November 19.

———— 2002. *What Went Wrong? Western Impact and Middle Eastern Response.* Oxford: Oxford University Press.

Lewis, Iain M. 1984. Sufism in Somaliland: A Study in Tribal Islam. In Ahmed S. Akbar and David M. Hart, editors, *Islam in Tribal Societies: From the Atlas to the Indus*, 127–168. London: Routledge and Kegan Paul.

Lindholm, Charles 1995. The New Middle Eastern Ethnography. *Journal of the Royal Anthropological Institute* n.s. 1: 805–820.

———— 1996. *The Islamic Middle East: An Historical Anthropology.* Oxford: Blackwell Publishers.

Lukens-Bull, Ronald A. 1999. Between Texts and Practice: Considerations in the Anthropology of Islam. *Marburg Journal of Religion* 4(2): 1–10.

———— 2001. Two Sides of the Same Coin: Modernity and Tradition in Islamic Education in Indonesia. *Anthropology & Education Quarterly* 32: 350–372.

Mabro, Judy 1991. *Veiled Half-Truths: Western Travellers' Perceptions of Middle Eastern Women.* London: I. B. Taurus & Co.

Magnarella, Paul 1974. *Tradition and Change in a Turkish Town.* Cambridge: Schenkman.

Mahdi, Muhsin 1957. *Ibn Khaldun's Philosophy of History.* London: George Allen & Unwin.

Mahmoud, Saba 2001a. Feminist Theory, Embodiment, and the Docile Agent: Some Reflections on the Egyptian Islamic Revival. *Cultural Anthropology* 16(2): 202–236.

———— 2001b. Rehearsed Spontaneity and the Conventionality of Ritual: Disciplines of Salat. *American Ethnologist* 28(4): 827–853.

Majid, Anouar 2000. *Unveiling Traditions: Postcolonial Islam in a Polycentric World.* Durham: Duke University Press.

Makdisi, George 1989. Scholasticism and Humanism in Classical Islam and the Christian West. *Journal of the American Oriental Society* 109(2): 175–182.

Malti-Douglas, Fedwa 1991. *Woman's Body, Woman's Word: Gender and Discourse in Arabo-Islamic Writing.* Princeton: Princeton University Press.

Mamdani, Mahmood 2002. Good Muslim, Bad Muslim: A Political Perspective on Culture and Terrorism. *American Anthropologist* 104(3): 766–775.

Maroof, S. M. 1981. Elements for an Islamic Anthropology. In Ismail R. Al-Faruqi and Abdulla Omar Naseef, editors, *Social and Natural Sciences: Islamic Education Series.* London: Hodder and Stoughton.

Martin, Richard C. 1984. Clifford Geertz Observed: Understanding Islam as Cultural Symbolism. In R. L. Moore and F. E. Reynolds, editors, *Anthropology and the Study of Religion*, 11–30. Chicago: Center for the Scientific Study of Religion.

———— 1985. Islam and Religious Studies: An Introductory Essay. In Richard C. Martin, editor, *Approaches to Islam in Religious Studies*, 1–21. Tucson: University of Arizona Press.

———— 1996. *Islamic Studies: A History of Religions Approach.* Upper Saddle River: Prentice Hall.

Martin, Richard C., Mark R. Woodward, Dwi S. Atmaja 1997. *Defenders of Reason in Islam: Mu'tazilism from Medieval School to Modern Symbol.* Oxford: Oneworld.

Ma'ruf, Muhammad 1989. Western Anthropology: A Critique of Evolutionism. In Mahmoud Abu Saud et al., *Toward Islamization of Disciplines,* 165–195. Herndon, VA: International Institute of Islamic Thought.

Marx, Emmanuel 1967. *The Bedouin of the Negev.* Manchester: Manchester University Press.

Mernissi, Fatima 1982. Virginity and Patriarchy, *Women's Studies International Forum* 5(2): 183–191.

———— 1986. *Women in Moslem Paradise.* New Delhi: Kale for Women.

———— 1987[1975]. *Beyond the Veil: Male-female Dynamics in Modern Muslim Society.* Revised edition. Bloomington: Indiana University Press.

———— 1991. *The Veil and the Male Elite: A Feminist Interpretation of Women's Rights in Islam.* Translated by Mary Jo Lakeland. Reading: Addison-Wesley.

———— 1994. *Dreams of Trespass: Tales of a Harem Girlhood.* Reading: Addison-Wesley.

Messick, Brinkley 1993. *The Calligraphic State: Textual Domination and History in a Muslim Society.* Berkeley: University of California Press.

Metcalf, Barbara D. 1989. *Discovering Islam; Making Sense of Muslim History and Society.* Akbar S. Ahmed. *Pacific Affairs* 62(1): 90–92 (Review).

Minh-ha, Trinh T. 1989. *Women, Native, Other.* Bloomington: Indiana University Press.

Mir-Hosseini, Ziba 1999. *Islam and Gender: The Religious Debate in Contemporary Iran.* Princeton: Princeton University Press.

Mohanty, Chandra Talpade 1994. Under Western Eyes: Feminist Scholarship and Colonial Discourse. In P. Williams and L. Chrisman, editors, *Colonial Discourse and Post-Colonial Theory: A Reader,* 196–220. New York: Columbia University Press (original, 1988).

Morey, Robert 1992. *The Islamic Invasion: Confronting the World's Fastest Growing Religion.* Eugene, OR: Harvest House Publishers.

Moosa, Ebrahim 2003. The Debts and Burdens of Critical Islam. In Omid Safi, editor, *Progressive Muslims on Justice, Gender and Pluralism,* 111–127. Oxford: Oneworld.

Moseley, C. W. R. D., translator 1983. *The Travels of Sir John Mandeville.* Middlesex: Penguin Books.

Muhammad, Mahathir 1989. Islamization of Knowledge and the Future of the Ummah. In Abu Saud et al., *Toward Islamization of Disciplines,* 19–24. Herndon, VA: International Institute of Islamic Thought.

Muir, William, 1912. *The Life of Mohammad.* Edinburgh: J. Grant.

Munson, Henry Jr. 1986. Geertz on Religion: The Theory and the Practice. *Religion* 16: 19–32.

———— 1987. *Islam and Revolution in the Middle East.* New Haven: Yale University Press.

Munson, Henry Jr. 1993a. *Religion and Power in Morocco.* New Haven: Yale University Press.

———— 1993b. Rethinking Gellner's Segmentary Analysis of Morocco's Ait 'Atta. *Man* (n.s.) 28: 267–280.

———— 1995. Segmentation: Reality or Myth? *The Journal of the Royal Anthropological Institute* 1(4): 821–829.

Murata, Sachiko 1992. *The Tao of Islam: A Sourcebook on Gender Relationships in Islamic Thought.* Albany: SUNY Press.

Murphy, Richard M. 2000. The Hairbrush and the Dagger: Mediating Modernity in Lahore. In Walter Armbrust, editor, *Mass Mediations: New Approaches to Popular Culture in the Middle East and Beyond*, 203–223. Berkeley: University of California Press.

Myers, Eugene A. 1963. *Arabic Thought and the Western World.* New York: Ungar.

Nagata, Judith 1982. Islamic Revival and the Problem of Legitimacy among Rural Religious Elites in Malaysia. *Man* (n.s.) 17: 42–57.

Needham, Rodney 1972. *Belief, Language, and Experience.* Chicago: University of Chicago Press.

Nissim-Sabat, Charles 1987. On Clifford Geertz and his "Anti Anti-Relativism." *American Anthropologist* 89(4): 935–939.

Norton, William Harmon 1924. The Influence of the Desert on Early Islam. *Journal of Religion* 4: 383–396.

Osman, Ghada 2003. The Historian on Language: Ibn Khaldun and the Communicative Learning Approach. *MESA Bulletin* 37(1): 50–57.

Oxford English Dictionary (OED)

Pals, Daniel L. 1996. *Seven Theories of Religion.* Oxford: Oxford University Press.

Peacock, James 1978. *Purifying the Faith: The Muhammadiyah Movement in Indonesian Islam.* Menlo Park: Cummings.

Peters, Emyrs 1960. The Proliferation of Segments in the Lineage of the Bedouin of Cyrenaica. *Journal of the Royal Anthropological Institute* 90: 29–53.

———— 1966. Preface. In W. Robertson Smith, *Kinship and Marriage in Early Arabia*, iii–xiii. Boston: Beacon Press.

———— 1967. Some Structural Aspects of the Feud among the Camel-Herding Bedouin of Cyrenaica. *Africa* 37(3): 261–281.

———— 1984. The Paucity of Ritual among Middle Eastern Pastoralists. In Ahmed S. Akbar and David M. Hart, editors, *Islam in Tribal Societies: From the Atlas to the Indus*, 187–219. London: Routledge and Kegan Paul.

Peterson, Houston, editor 1958. *Essays in Philosophy from David Hume to Bertrand Russell.* New York: Washington Square Press.

Prakash, Buddha 1954. Ibn Khaldun's Philosophy of History. *Islamic Culture* 28: 492–508.

Prideaux, Humphrey 1697. *The True Nature of Imposture Fully Displayed in the Life of Mahomet.* London: William Rogers.

Rabinow, Paul 1977. *Reflections on Fieldwork in Morocco.* Berkeley: University of California Press.

Rahman, Afzalur, editor 1986. *Muhammad Encyclopaedia of Seerah.* Vol. 2. London: Seerah Foundation.

Rahman, Fazlur 1968. *Islam.* Garden City: Doubleday Anchor Books.

Redfield, Robert 1960. *Peasant Society and Culture.* Chicago: The University of Chicago Press.

Reiss, Edmund 1967. *Elements of Literary Analysis.* Cleveland: The World Publishing Company.

Robbins, Joel 2003. What is a Christian? Notes toward an Anthropology of Christianity. *Religion* 33: 191–199.

Roberts, Hugh 2002. Perspectives on Berber Politics: On Gellner and Masqueray, or Durkheim's Mistake. *Journal of the Royal Anthropological Institute* 8: 107–126.

Roff, William R. 1985. Islam Obscured? Some Reflections on Studies of Islam and Society in Southeast Asia. *Archipel* 29: 7–34.

Rosen, Lawrence 2002. *The Culture of Islam: Changing Aspects of Contemporary Muslim Life.* Chicago: University of Chicago Press.

Rosenthal, Erwin 1932. *Ibn Khalduns Gedanken uber den Staat, ein Beitrag zur Geschichte der mittelalterlichen Staatslehre.* Munich: R. Oldenbourg.

Rosenthal, Franz 1958. *Political Thought in Medieval Islam; An Introductory Outline.* Cambridge: Cambridge University Press.

Safi, Omid 2003. Introduction: the Times They Are A-Changin'—A Muslim Quest for Justice, Gender Equality, and Pluralism. In Omid Safi, editor, *Progressive Muslims on Justice, Gender and Pluralism,* 1–29. Oxford: Oneworld.

Said, Edward 1979. *Orientalism.* New York: Vintage Books.

——— 1982. Orientalism: An Exchange. *The New York Review of Books* 29(13): 44–46 [June 24].

——— 1986. Foucault and the Imagination of Power. In David C. Hoy, editor, *Foucault: A Critical Reader,* 149–155. Oxford: Blackwell.

——— 1994. Afterword. In *Orientalism,* 329–352. New York: Vintage Books.

Salzman, Philip 1975. Islam and Authority in Tribal Iran: A Comparative Comment. *The Moslem World* 55: 186–195.

Schick, Irvin Cemil 1999. *The Erotic Margin: Sexuality and Spatiality in Alterist Discourse.* London: Verso.

Schimmel, Annemarie 1955. Zur Anthropologie des Islam. In C. J. Bleeker, editor, *L'anthropologie religieuse: L'homme et sa destinée à la lumière de l'histoire des religions.* Studies in the History of Religions. Supplement of *Numen,* II, 140–154. Leiden: Brill.

——— 1985. *And Muhammad Is His Messenger: The Veneration of the Prophet in Islamic Piety.* Durham: University of North Carolina Press.

Schlegel, Stuart A. 1998. *Wisdom from a Rainforest: The Spiritual Journey of an Anthropologist.* Athens: The University of Georgia Press.

Schmidt, Nathaniel 1930. *Ibn Khaldun: Historian, Sociologist and Philosopher.* New York: Columbia University Press.

Schmidt, Wilhelm 1962ff. *Der Ursprung der Gottesidee.* Münster: Aschendorffsche Verlagsbuchhandlung.

Schneider, Jane and Peter 2002. The Mafia and al-Qaeda: Violent and Secretive Organisations in Comparative and Historical Perspective. *American Anthropologist* 104(3): 776–782.

Scupin, Raymond 2000. The Anthropological Perspective on Religion. In Raymond Scupin, editor, *Religion and Culture: An Anthropological Focus*, 1–15. Upper Saddle River, NJ: Prentice Hall.

Sells, Michael 1999. *Approaching the Qur'ân: The Early Revelations.* Ashland: White Cloud Press.

——— 2002. Understanding, Not Indoctrination. *The Washington Post*, August 8, 2002, A17.

Serjeant, Robert Bertram 1982. The Interplay between Tribal Affinities and Religious (Zaydi) Authority in the Yemen. *al-Abhath* 30: 11–50.

Sewell, William H. Jr. 1997. Geertz, Cultural Systems, and History: From Synchrony to Transformation. *Representations* 59: 35–55.

Shahrani, Nazif 2002. War, Factionalism, and the State in Afghanistan. *American Anthropologist* 104(3): 715–722.

Shahrani, Nazif and Robert L. Canfield, editors. 1984. *Revolutions and Rebellions in Afghanistan: Anthropological Perspectives.* Berkeley: Institute of International Studies.

Shamsul, A. B. 1997. Identity Construction, Nation Formation, and Islamic Revivalism in Malaysia. In Robert W. Hefner and Patricia Horvatich, editors, *Islam in an Era of Nation-States: Politics and Religious Renewal in Muslim Southeast Asia*, 207–227. Honolulu: University of Hawaii Press.

Shankman, Paul 1984. The Thick and the Thin: On the Interpretive Theoretical Program of Clifford Geertz. *Current Anthropology* 25: 261–270.

Sharabi, Hisham 1990. The Scholarly Point of View: Politics, Perspective, Paradigm. In Hisham Sharabi, editor, *Theory, Politics and the Arab World: Critical Responses*, 1–51. London: Routledge.

Shariati, Ali 1979. *On the Sociology of Islam.* Edited by Hamid Algar. Berkeley: Mizan.

Shryock, Andrew 2002. New Images of Arab Detroit: Seeing Otherness and Identity through the Lens of September 11. *American Anthropologist* 104(3): 917–922.

Siegel, Robert A. 1999. Weber and Geertz on the Meaning of Religion. *Religion* 29: 61–71.

Sim, Stuart, editor 2001. *The Routledge Companion to Postmodernism.* London: Routledge.

Sivan, Emmanuel 1985. *Interpretations of Islam: Past and Present.* Princeton: The Darwin Press.

Smith, Jonathan Z. 1978. I am a Parrot (Red). In *Map is Not Territory*, 265–298. Chicago: University of Chicago Press (original, 1972).

Smith, Wilfred Cantwell 1957. *Islam in Modern History.* New York: New American Library.

Southern, Richard 1962. *Western Views of Islam in the Middle Ages.* Cambridge: Harvard University Press.

Spellberg, Denise 1994. *Politics, Gender, and the Islamic Past: The Legacy of 'Aisha bint Abi Bakr.* New York: Columbia University Press.

Spencer, Jonathan 1989. Anthropology as a Kind of Writing. *Man* (n.s.) 24(1): 145–164.

Spickard, James V. 2001. Tribes and Cities: Towards an Islamic Sociology of Religion. *Social Compass* 48(1): 103–116.

Spiegel, Gabrielle M. 1997. *The Past as Text: The Theory and Practice of Medieval Historiography.* Baltimore: Johns Hopkins Press.

Starrett, Gregory 1995a. The Political Economy of Religious Commodities in Cairo. *American Anthropologist* 97(1): 51–68.

——— 1995b. The Hexis of Interpretation: Islam and the Body in the Egyptian Popular School. *American Ethnologist* 22(4): 953–969.

——— 1998. *Putting Islam to Work: Education, Politics, and Religious Transformation in Egypt.* Berkeley: University of California Press.

Stirling, Paul 1965. *Turkish Village.* New York: John Wiley and Sons.

Stowasser, Barbara Freyer 1993. Women's Issues in Modern Islamic Thought. In J. E. Tucker, editor, *Arab Women: Old Boundaries, New Frontiers,* 3–28. Bloomington: Indiana University Press.

——— 1994. *Women in the Qur'an, Traditions, and Interpretation.* Oxford: Oxford University Press.

Strathern, Marilyn 1987. An Awkward Relationship: The Case of Feminism and Anthropology. *Signs* 12(2): 276–293.

Strijp, Ruud 1992. *Cultural Anthropology of the Middle East: A Bibliography.* Handbuch der Orientalistik, 1. Abteilung. Der Nähe und der Mitterlere Osten, Band 10. Leiden: Brill.

Sweet, Louise, editor 1970. *Peoples and Cultures of the Middle East: An Anthropological Reader.* Garden City: Natural History Press.

——— 1974. *Tell Toqaan: A Syrian Village.* Museum of Anthropology Anthropological Papers, 14. Ann Arbor: University of Michigan.

Swenson, J. D. 1985. Martyrdom: Mytho-cathexis and the Mobilization of the Masses in the Iranian Revolution. *Ethos* 13(2): 121–149.

Talbi, M. 1971. IBN KHALDÛN, *The Encyclopedia of Islam,* 2nd ed. 3: 825–831. Leiden: Brill.

Tapper, Nancy 1979. Mysteries of the Harem? An Anthropological Perspective on Recent Studies of Women of the Muslim Middle East, *Women's Studies International Quarterly* 2: 481–487.

——— 1991. *Bartered Brides: Politics, Gender, and Marriage in an Afghan Tribal Society.* Cambridge: Cambridge University Press.

Tapper, Nancy and Richard 1987. The Birth of the Prophet: Ritual and Gender in Turkish Islam, *Man* (n.s.) 22: 69–92.

Tapper, Richard 1984. Holier than Thou: Islam in Three Tribal Societies. In Akbar S. Ahmed and David M. Hart, editors, *Islam in Tribal Societies: From the Atlas to the Indus,* 244–265. London: Routledge and Kegan Paul.

Tapper, Richard 1995. "Islamic Anthropology" and the "Anthropology of Islam." *Anthropological Quarterly* 68(3): 185–193.

Thomson, William M. 1901. *The Land and the Book or, Biblical Illustrations Drawn from the Manners and Customs, the Scenes and Scenery of the Holy Land*. London: T. Nelson and Sons (original, 1858).

Thomson, William 1933. Erwin Rosenthal, *Ibn Khalduns Gedanken uber den Staat, ein Beitrag zur Geschichte der mittelalterlichen Staatslehre...* *Speculum* 8:1:109–113 (Review).

Tibi, Bassam 2001. *Islam between Culture and Politics*. New York: Palgrave.

Titus, Paul 1995. Islam and the Evil Demon. *American Anthropologist* 97(3): 543–558.

Toth, James 2003. Islamism in Southern Egypt: A Case Study of a Radical Religious Movement. *International Journal of Middle East Studies* 35: 547–572.

Toynbee, Arnold 1935. *A Study of History* III. London: Oxford University Press.

Trencher, Susan 2000. *Mirrored Images: American Anthropology and American Culture, 1960–1980*. Westport: Bergin & Garvey.

Tucker, Judith E. 1993. Introduction. In her *Arab Women: Old Boundaries, New Frontiers*, vii–xviii. Bloomington: Indiana University Press.

Turner, Bryan S. 1994. *Orientalism, Postmodernism and Globalism*. London: Routledge.

——— 2003. *Islam: Critical Concepts in Sociology*. 7 volumes. London: Routledge [Editor].

Tylor, Edward 1871. *Primitive Culture*. London: Murray.

——— 1881. *Anthropology: An Introduction to the Study of Man and Civilization*. London: Macmillan.

Van-Lennep, Henry J. 1875. *Bible Lands: Their Modern Customs and Manners Illustrative of Scripture*. New York: Harper & Brothers.

van Nieuwenhuijze, C. A. O. 1979. Jean-Paul CHARNAY, *Sociologie religieuse de l'Islam, Préliminaires... Bibliotheca Orientalis* 36(5–6): 383–385 (Review).

Varisco, Daniel Martin 1995. Metaphors and Sacred History: The Genealogy of Muhammad and the Arab "Tribe." *Anthropological Quarterly* 68(3): 139–156.

——— 2000. Slamming Islam: Participant Webservation with a Web of Meanings to Boot. *Working Papers of the MES*. Electronic Document, http://www.aaanet.org/mes/lectvar1.htm. Accessed April, 2004.

——— 2002a. The Archaeologist's Spade and the Apologist's Stacked Deck: The Near East through Conservative Christian Bibliolatry. In Abbas Amanat and Magnus T. Bernhardssen, editors, *The United States & the Middle East: Cultural Encounters*, 57–113. New Haven: The Yale Center for International and Area Studies.

——— 2002b. September 11: Participant Webservation of the "War on Terrrorism." *American Anthropologist* 104(3): 934–938.

——— 2004. Reading Against Culture in Edward Said's *Culture and Imperialism*. *Culture, Theory and Critique* 45(2): 1–20.

von Grunebaum, Gustave E. 1954. *Medieval Islam: A Study in Cultural Orientation*, 2nd ed. Chicago: University of Chicago Press.
———— 1961. An Analysis of Islamic Civilization and Cultural Anthropology. In *Colloque sur la Sociologie Musulmane*, 21–73. Bruxelles.
Wadud, Amina 1999. *Qur'an and Woman: Rereading the Sacred Text from a Woman's Perspective*. Oxford: Oxford University Press.
Waghorne, Joanne Punzo 1984. From Geertz's Ethnography to an Ethno-theology? In R. L. Moore and F. E. Reynolds, editors, *Anthropology and the Study of Religion*, 31–55. Chicago: Center for the Scientific Study of Religion.
Waldman, Marilyn R. 1985. Primitive Mind/Modern Mind: New Approaches to an Old Problem Applied to Islam. In Richard C. Martin, editor, *Approaches to Islam in Religious Studies*, 91–105. Tucson: University of Arizona Press.
Wallace, Anthony 1966. *Religion: An Anthropological View*. New York: Random House.
Walters, Ronald 1980. Signs of the Times: Clifford Geertz and the Historians. *Social Research* 47: 536–556.
Weber, Max 1964. *The Sociology of Religion*. Translated by Ephraim Fischoff. Boston: Beacon Press.
Weiss, Dieter 1995. Ibn Khaldun on Economic Transformation, *International Journal of Middle East Studies* 27: 29–37.
Werbner, Pnina 1996. Allegories of Sacred Imperfection: Magic, Hermeneutics, and Passion in The Satanic Verses. *Current Anthropology*, Supplement: Special Issue: Anthropology in Public, 37(1): S55–S86.
Westermarck, Edward 1916. *The Moorish Conception of Holiness (Baraka)* Öfversigt af Finska Vetenskapo-Societetens Förhandlingar, 57, B, 1. Helsingfors.
———— 1926. *Ritual and Beliefs in Morocco*. 2 volumes. London: Macmillan.
White, Jenny B. 1999. Amplifying Trust: Community and Communication in Turkey. In Dale F. Eickelman and Jon W. Anderson, editors, *New Media in the Muslim World: The Emerging Public Sphere*, 162–179. Bloomington: Indiana University Press.
———— 2002. *Islamist Mobilization in Turkey: A Study in Vernacular Politics.* Seattle: University of Washington Press.
Williams, Bernard 1973. Hume on Religion. In D. F. Pears, editor, *David Hume: A Symposium*. London: Macmillan.
Wolf, Eric 1951. The Social Organization of Mecca and the Origins of Islam. *Southwestern Journal of Anthropology* 7: 329–356.
Woodward, Mark R. 1989. *Islam in Java: Normative Pietry and Mysticism in the Sultanate of Yogyakarta*. Tucson: University of Arizona Press.
———— 1996. Introduction. Talking across Paradigms: Indonesia, Islam, and Orientalism. In Mark R. Woodward, editor, *Toward a New Paradigm: Recent Developments in Indonesian Islamic Thought*, 1–45. Tempe: Arizona State University, Program for Southeast Asian Studies.
Young, William C. 1993. The Ka'ba, Gender, and the Rites of Pilgrimage. *International Journal of Middle East Studies* 25: 285–300.

Young, William C. 1996. *The Rashayyda Bedouin: Arab Pastoralists of Eastern Sudan*. Fort Worth: Harcourt Brace College Publishers.

Yeazell, Ruth Bernard 2000. *Harems of the Mind: Passages of Western Art and Literature*. New Haven: Yale University Press.

Zammito, John H. 2002. *Kant, Herder, and the Birth of Anthropology*. Chicago: University of Chicago Press.

Index

ANTHROPOLOGY

"In *Islam Obscured* Daniel M. Varisco offers a brilliant and nuanced analysis of four influential anthropologists against the background of older, less field-based ethnographies of Muslim societies. His penetrating critiques of the influential works of Clifford Geertz, Ernest Gellner, Fatima Mernissi, and Akbar Ahmed lead to a probing discussion of the challenges facing the anthropology of Islam in the twenty-first century."

—RICHARD C. MARTIN, PROFESSOR OF ISLAMIC STUDIES AND HISTORY OF RELIGIONS, EMORY UNIVERSITY

Islam Obscured analyzes four seminal anthropology texts on Muslims that have been read widely outside the discipline. Two are by distinguished anthropologists: *Islam Observed* (Clifford Geertz, 1968) and *Muslim Society* (Ernest Gellner, 1981). Two other texts are by Muslim scholars: *Beyond the Veil* (Fatima Mernissi, 1975) and *Discovering Islam* (Akbar Ahmed, 1988). Varisco argues that each of these approaches Islam as an essentialized organic unity rather than letting "islams" found in the field speak to the diversity of practice. He sheds light on Islam as a cultural phenomenon, representation of the other, Muslim gender roles, politics of ethnographic authority, and Orientalist discourse. Varisco's analysis goes beyond the rhetoric over what Islam is, focusing instead on ethnographic research about what Muslims say they do and actually are observed doing.

DANIEL MARTIN VARISCO is chair of the Department of Anthropology, Hofstra University.

Cover photo by Henry Bechard, ca. 1870.
"Interior of the Amron Mosque"
in a private collection.
Cover design by Julia Kushnirsky

ISBN 1-4039-6773-3

9 781403 967732

www.palgrave.com